GENETICALLY MODIFIED PEST-PROTECTED PLANTS

SCIENCE AND REGULATION

Committee on Genetically Modified Pest-Protected Plants

Board on Agriculture and Natural Resources

National Research Council

NATIONAL ACADEMY PRESS
Washington, D.C.

NATIONAL ACADEMY PRESS • 2101 Constitution Avenue, NW • Washington, DC 20418

NOTICE: The project that is the subject of this report was approved by the Governing Board of the National Research Council, whose members are drawn from the councils of the National Academy of Sciences, the National Academy of Engineering, and the Institute of Medicine. The members of the committee responsible for the report were chosen for their special competences and with regard for appropriate balance.

Library of Congress Cataloging-in-Publication Data

Genetically modified pest-protected plants : science and regulation / Committee on Genetically Modified Pest-Protected Plants, Board on Agriculture and Natural Resources, National Research Council.
    p. cm.
Includes bibliographical references and index.
  ISBN 0-309-06930-0 (casebound)
  1. Transgenic plants—Risk assessment. 2. Plants—Disease and pest resistance—Genetic aspects. I. National Research Council (U.S.). Committee on Genetically Modified Pest-Protected Plants.
  SB123.57 .G48 2000
  631.5'233—dc21
                              00-009457

Genetically Modified Pest-Protected Plants: Science and Regulation is available from National Academy Press, 2101 Constitution Avenue, NW, Lockbox 285, Washington, DC 20055; (800) 624-6242 or (202) 334-3313 (in the Washington metropolitan area); Internet, http://www.nap.edu

THE NATIONAL ACADEMIES

National Academy of Sciences
National Academy of Engineering
Institute of Medicine
National Research Council

The **National Academy of Sciences** is a private, nonprofit, self-perpetuating society of distinguished scholars engaged in scientific and engineering research, dedicated to the furtherance of science and technology and to their use for the general welfare. Upon the authority of the charter granted to it by the Congress in 1863, the Academy has a mandate that requires it to advise the federal government on scientific and technical matters. Dr. Bruce M. Alberts is president of the National Academy of Sciences.

The **National Academy of Engineering** was established in 1964, under the charter of the National Academy of Sciences, as a parallel organization of outstanding engineers. It is autonomous in its administration and in the selection of its members, sharing with the National Academy of Sciences the responsibility for advising the federal government. The National Academy of Engineering also sponsors engineering programs aimed at meeting national needs, encourages education and research, and recognizes the superior achievements of engineers. Dr. William A. Wulf is president of the National Academy of Engineering.

The **Institute of Medicine** was established in 1970 by the National Academy of Sciences to secure the services of eminent members of appropriate professions in the examination of policy matters pertaining to the health of the public. The Institute acts under the responsibility given to the National Academy of Sciences by its congressional charter to be an adviser to the federal government and, upon its own initiative, to identify issues of medical care, research, and education. Dr. Kenneth I. Shine is president of the Institute of Medicine.

The **National Research Council** was organized by the National Academy of Sciences in 1916 to associate the broad community of science and technology with the Academy's purposes of furthering knowledge and advising the federal government. Functioning in accordance with general policies determined by the Academy, the Council has become the principal operating agency of both the National Academy of Sciences and the National Academy of Engineering in providing services to the government, the public, and the scientific and engineering communities. The Council is administered jointly by both Academies and the Institute of Medicine. Dr. Bruce M. Alberts and Dr. William A. Wulf are chairman and vice chairman, respectively, of the National Research Council.

COMMITTEE ON GENETICALLY MODIFIED PEST-PROTECTED PLANTS

PERRY ADKISSON, *Chair*, Texas A&M University, College Station
STANLEY ABRAMSON, Arent, Fox, Kintner, Plotkin & Kahn, Washington, D.C.
STEPHEN BAENZIGER, University of Nebraska, Lincoln
FRED BETZ, Jellinek, Schwartz & Connolly, Arlington, Virginia
JAMES C. CARRINGTON, Washington State University, Pullman
REBECCA J. GOLDBURG, Environmental Defense, New York, NY
FRED GOULD, North Carolina State University, Raleigh
ERNEST HODGSON, North Carolina State University, Raleigh
TOBI JONES, California Department of Pesticide Regulation, Sacramento
MORRIS LEVIN, University of Maryland, Baltimore
ERIK LICHTENBERG, University of Maryland, College Park
ALLISON SNOW, Ohio State University, Columbus

Staff

JENNIFER KUZMA, Study Director
MICHAEL J. PHILLIPS, Study Director (through July 1999)*
JAMIE YOUNG, Research Associate
KAREN L. IMHOF, Project Assistant
DEREK SWEATT, Project Assistant
NORMAN GROSSBLATT, Editor

---

*Michael Phillips was involved with this study until 7/13/99 and is currently employed with the Biotechnology Industry Organization.

Foreword

THE CONTEXT OF THIS REPORT

A revolution has been taking place in the life sciences, sparked by striking advances in our fundamental understanding of living systems. These advances have led to the development of powerful molecular techniques, which can help society to conquer human disease, improve food production, and better protect the environment. As with all new scientific developments, however, potential risks need to be carefully evaluated and dealt with appropriately. The National Academies are committed to bringing together experts to discuss and comment on the scientific issues surrounding the application of biotechnology to important modern-day problems.

In 1987 the National Academy of Sciences issued a white paper on the "Introduction of Recombinant DNA-Engineered Organisms into the Environment," which dealt with general principles concerning potential ecological risks in field testing. In his preface, my predecessor, Frank Press, stated that the paper "applies the relevant scientific principles" to key issues, but was not intended to "resolve questions pertaining to the establishment of specific regulations or guidelines governing release procedures." In 1989, the National Research Council issued the report, "Field Testing Genetically Modified Organisms: Framework for Decisions," which addressed the ecological risks of small-scale field testing of engineered organisms. Neither potential human health risks, nor issues raised by large-scale commercial planting, were addressed in that study which considered scientific issues primarily, not regulatory policy. These two reports reflected the best judgment of two highly expert groups of scientists, and they were based on the scientific evidence available to them at the time. The full text of these and all other reports from the National Academies are available on the Worldwide Web at www.nap.edu.

Utilizing information gained over the past decade, the National Research Council is releasing this important report on genetically modified pest-protected plants. Prepared by another expert committee, it provides timely advice to researchers, developers, and regulatory agencies involved in reviewing the science surrounding the regulation of genetically modified pest-protected crops. The report addresses only one aspect of the ongoing revolution in the life sciences and agriculture, and it is careful to point out where more research and scientific information is needed to answer remaining questions. The National Research Council intends for it to be only the first of several reports to be produced over the next couple of years. We have recently established a standing committee on Biotechnology, Food and Fiber Production, and the Environment. This committee will oversee a wide range of studies, workshops, and meetings. In this way, we look forward to being able to contribute on an ongoing basis to discussions of the important and timely issues surrounding agricultural biotechnology.

PROTECTING PLANTS FROM PESTS

Agriculture has been suffering from pest and disease infestation since its inception, causing enormous, unpredictable losses in food production. Genetic engineering of plants for resistance to pests and disease, creating transgenic pest-protected plants, is one of the many tools for increasing food security. It is embedded within the long-standing science of conventional breeding for plant improvement. The use of chemicals to control pests[1] can be abated and perhaps someday eliminated by the appropriate use of transgenic methods, combined with conventional plant breeding and other techniques of sustainable agriculture.

Many valuable technologies will form the basis for future plant protection. The appropriate balance among them will be pest- and situation-specific. Given time constraints, this report does not include an in-depth

---

[1]The forthcoming NRC report *The Future Role of Pesticides in U.S. Agriculture*, will deal with the use of chemicals as a trend in pest management.

analysis of this balance.[2] It instead provides an overview of the use of transgenic techniques to enhance the pest resistance of crops, with a focus on the regulatory system that oversees the introduction of transgenic pest-protected plants. In this sense, it is but one contribution to the larger and complex system of pest management, as well as to the broader issues surrounding the often virulent debate about using modern biotechnology to improve agricultural production.

THE PREPARATION OF THIS REPORT

In the preparation of this report, much effort was placed on selecting highly qualified experts capable of addressing the scientific and regulatory issues surrounding the regulation of genetically modified pest-protected plants. Care was also given to achieving an appropriate balance of viewpoints. Suggestions for committee members came from many different sources, including extensive public comments. This report represents the consensus views of the 12 experts who were selected by the National Research Council to undertake the study.

Care was also given to ensuring that the committee received input and information from all concerned and interested parties. A public workshop was held in which the public and many panelists from diverse perspectives were invited to express their ideas and concerns about transgenic pest-protected plants and the regulatory framework guiding their commercial use. The committee's analysis utilized input from the workshop, as well as from a variety of other scientific sources.

Although funded entirely with internal funds from the National Academies, the public disclosure procedures of Section 15 of the Federal Advisory Committee Act were used to guide the committee process. Committee membership and public workshops were posted on the Web on our Current Projects system. As with all NRC studies, this report has been subject to an extensive independent peer review. Twelve scientific and regulatory experts, representing a broad range of viewpoints, reviewed the report and provided extensive comments, and they thereby helped the committee to strengthen the report.

---

[2]The 1996 NRC report, *Ecologically Based Pest Management – New Solutions for a New Century* provides an overview of the management of the myriad biological processes that suppress pest buildup and damage and of the increasing contributions of production ecology to the future of agriculture. Available online at http://books.nap.edu/catalog/5135.html.

ACKNOWLEDGMENTS

I would like to thank Dr. Perry Adkisson, the committee chair, and the 11 other committee members for their dedicated, *pro bono* work on this study. Special thanks are also due to Dr. Jennifer Kuzma, who took over last July as the NRC Responsible Staff Officer for this report, early in its preparation.

Bruce Alberts
Chairman
National Research Council

Preface

Transgenic pest-protected crops were first commercially planted in the United States in 1995. Since then the acreage planted to transgenic crops has increased rapidly with some 70 million acres being grown in the United States, and 98.6 globally in 1999. Of this acreage, a large percentage (for example, 30 million acres in the US in 1999) is planted with transgenic pest-protected crop varieties containing the *Bacillus thuringiensis (Bt)* gene which confers protection from certain insect pests and with varieties that are herbicide-tolerant. In 1998, about 25% of the US cotton acreage and 21% of the corn acreage was planted with varieties containing *Bt* genes.

This increase in acreage planted in transgenic crops has largely resulted because of benefits produced to farmers. Many farmers are growing transgenic crops because they either produce more effective control of serious pests than conventional chemical treatments, or they provide control at lower costs than conventional treatments, or both. The growing of some *Bt* crops has been accompanied by a reduction in the amounts of chemical pesticides previously used on these crops. This has produced a side benefit in terms of reducing exposure of humans and other non-target organisms to these toxic chemicals and lessening the contamination of air and water.

Given the rapid increase in plantings of transgenic varieties, concerns have been raised about the ecological and human health risks that might be posed by these crops. Although these risks might not in principle differ in type from those associated with other conventionally-bred pest-resistant varieties or chemical pesticides, they nevertheless have become a focus of attention by several groups who are concerned by potential risks that might be posed by transgenic breeding methods. This concern has been magnified in Europe and other parts of the world where consumer resistance has been increasing against food products produced from transgenic plants.

Concerns about the risks posed by transgenic plants have led some to question the safety review they receive in the United States under the Coordinated Regulatory Framework. Some believe that human health and environmental risks are not properly assessed. Others believe the risks are minimal, that benefits outweigh risks, and the current regulatory scheme is too onerous. This debate has intensified in recent months given the international climate and impending regulatory decisions in the United States where new regulations for transgenic plants are being considered.

Several professional societies, members of Congress, and other groups have expressed concern over the regulation of transgenic crops, citing the need for an impartial review of the present and proposed process. The National Research Council responded to this need by commissioning and funding the present study which was initiated in March 1999. The committee was charged with the following task: "The Committee will investigate risks and benefits of genetically modified pest-protected (GMPP) plants and the coordinated Regulatory Framework for Regulation of Biotechnology affecting the use of these plants. The study will (1) review the principles in the NAS Council's white paper, *Introduction of Recombinant DNA-Engineered Organisms into the Environment* (1987), for their continued scientific validity and assess their appropriateness for current decisions regarding GMPP plants; (2) review scientific data which addresses the risks and benefits of GMPP plants; (3) examine the existing and proposed regulations to qualitatively assess their consequences for research, development, and commercialization of GMPP plants; and (4) provide recommendations to address the identified risk/benefits, and, if warranted, for the existing and proposed regulation of GMPP plants."

The committee was given a very short time frame and a limited budget for accomplishing this task. Committee members were identified in early spring 1999 and the first meeting was convened in April. Two later meetings followed this, one of which included a workshop in which public participation was invited. The meetings and the workshop provided the basis for the present report.

The report is composed of four chapters and an Executive Summary. Chapter 1 is an introductory chapter that discusses issues which led to the initiation of the present study, current EPA, USDA, and FDA policies, the task given to the committee by the NRC, and role of this report. Chapter 2 deals with the potential environmental and human health impacts of pest-protected plants with risks and benefits being among the issues discussed. Chapter 3 provides several case studies related to the commercial production of transgenic genetically modified pest-protected crops, analyzes the 1994 and 1997 rules proposed by EPA for the regulation of plant-pesticides, and identifies several research needs. Chapter 4 provides an overview of the current regulation of plant products under the coordinated framework for the regulation of biotechnology by EPA, FDA, and USDA and provides recommendations that the committee believes will improve this process. The Executive Summary summarizes the key finding, conclusions, and recommendations of the report.

Because the time-frame for the conduct of the present study was very short, there were several issues of public concern that were not included in our deliberations. For example, the committee did not consider issues involving herbicide-tolerant crops or labeling of food products produced from transgenic plants. The NRC's new Standing Committee on Biotechnology, Food and Fiber Production, and the Environment will be equipped to help to identify and examine many related issues in greater detail. Also, the committee gave more consideration to the potential risks posed by the commercialization of transgenic pest-protected plants than to benefits that they might produce to farmers and the environment.

In recent months there have been many reports in the mass media concerning the negative aspects of agricultural biotechnology. Little has been said about the positive impacts that transgenic plants are having on agricultural production and environmental quality. In the future, society and regulatory authorities must find a way to balance the risks and benefits of the use of this technology in the production of food and feed crops and develop appropriate processes for their regulation. As a committee we trust that the present report will help increase our knowledge of transgenic plants and our ability to make wiser decisions concerning their regulation.

Perry L. Adkisson
Chairman

Acknowledgments

The committee wishes to express its thanks to the staff members of the Board on Agriculture and Natural Resources for their invaluable assistance in the conduct of this study and the preparation of this report. Special thanks are due to Dr. Jennifer Kuzma, Project Director, for her dedicated efforts and hard work in compiling and assimilating the reports of the various subgroups of the committee and for shepherding the report through several reiterations to completion. The committee appreciates her technical competence in the conduct of the study and her diplomacy in resolving differences that arose during the writing of the report. The committee wishes to recognize the efforts of Dr. Michael Phillips who was study director of the project for the first four months.

The committee also wishes to recognize the outstanding work of Ms. Jamie Young, Research Associate, Ms. Karen Imhof, Project Assistant, and Mr. Derek Sweatt, Project Assistant, for their assistance in the work of the committee and preparation of this report. The committee appreciates the input of Dr. Jim Reisa, Director of the Board on Environmental Studies and Toxicology, in guiding project staff. The committee also acknowledges Mr. Norman Grossblatt, Editor, for his expert editorial assistance in improving the final draft of the report.

Special thanks are due tocommittee members Dr. Fred Gould and Mr. Stanley Abramson for assuming a large share of the workload of the committee by chairing the two technical subgroups that developed the bulk of the report.

The committee expresses their gratitude to the following people for the information they provided to the committee.   In some cases, the committee needed to obtain information on short notice, and the committee appreciates the efforts of these people to fulfill these requests.

Richard Allison, Michigan State University
Janet Andersen, Biopesticides and Pollution Prevention Division, Environmental Protection Agency
Nega Beru, Center for Food Safety and Applied Nutrition, Food and Drug Administration
Stacy Carey, House Agriculture Committee
Tom Carrato, Monsanto Company
Harold Coble, North Carolina State University
James Cook, Washington State University
Kent Croon, Monsanto Company
Tim Debus, United Fresh Fruit and Vegetable Association
Kathryn DiMatteo, Organic Trade Association
Steven Druker, Alliance for Biointegrity
Nina Fedoroff, Pennslyvania State University
David Ferro, University of Massachusetts at Amherst
Vasilios Frankos, Environ
Alan Goldhammer, Biotechnology Industry Organization
Dennis Gonsalves, Cornell University
Bob Harness, Monsanto Company
David Heron, Animal and Plant Health Inspection Service, United States Department of Agriculture
Jason Hlywka, University of Nebraska at Lincoln
Karen Hokanson, Animal and Plant Health Inspection Service, United States Department of Agriculture
Phil Hutton, Biopesticides and Pollution Prevention Division, Environmental Protection Agency
Peter Kareiva, Department of Commerce
John Kough, Biopesticides and Pollution Prevention Division, Environmental Protection Agency
Marc Lappe, Center for Ethics and Toxics
Nina Mani, George Washington University
James Maryanski, Center for Food Safety and Applied Nutrition, Food and Drug Administration
Sally McCammon, Animal and Plant Health Inspection Service, United States Department of Agriculture
Terry Medley, DuPont Company
Margaret Mellon, Union of Concerned Scientists

Mike Mendelsohn, Biopesticides and Pollution Prevention Division, Environmental Protection Agency
Robert Mustell, National Corn Growers Association
William Price, Center for Veterinary Medicine, Food and Drug Administration
Phil Regal, University of Minnesota
Marlin Rice, Iowa State University
Jennifer Riebe, Monsanto Company, NatureMark
Jane Rissler, Union of Concerned Scientists
Russ Schneider, Monsanto Company
Doreen Stabinsky, California State University
Guenther Stotzy, New York University
Gail Tomimatsu, Biopesticides and Pollution Prevention Division, Environmental Protection Agency
Robert Torla, Biopesticides and Pollution Prevention Division, Environmental Protection Agency
John Trumble, University of California, Riverside
Rick Welsh, Wallace Institute
James White, Animal and Plant Health Inspection Service, United States Department of Agriculture

Acknowledgment of Reviewers

This report has been reviewed in draft form by individuals chosen for their diverse perspectives and technical expertise in accordance with procedures for reviewing NRC reports approved by the NRC's Report Review Committee. The purpose of this independent review is to provide candid and critical comments that will assist the NRC in making the published report as sound as possible and to ensure that the report meets institutional standards for objectivity, evidence, and responsiveness to the study charge. The content of the final report is the responsibility of the NRC and the study committee, and not the responsibility of the reviewers. The review comments and draft manuscript remain confidential to protect the integrity of the deliberative process.

We wish to thank the following individuals, who are neither officials nor employees of the NRC, for their participation in the review of this report:

John Antle, Montana State University
John Benedict, Texas A&M University
Joy Bergelson, University of Chicago
Edwin Clark II, Washington, DC
John Dowling, Harvard University
Robert Fraley, Monsanto Company
Sarjeet Gill, University of California - Riverside
Lynn Goldman, Johns Hopkins University
Walter Goldstein, Michael Fields Agricultural Institute

Richard Harwood, Michigan State University
Susan Hefle, University of Nebraska at Lincoln
Jane Rissler, Union of Concerned Scientists
Jozef S. Schell, Max Planck Institute for Breeding Research
Luis Sequeira, University of Wisconsin

The individuals listed above have provided many constructive comments and suggestions. It must be emphasized, however, that responsibility for the final content of this report rests entirely with the authoring committee and the NRC.

Contents

APPENDIXES

Executive Summary

Pest and pathogen management to optimize crop health, productivity, food quality and safety is critical to global food security, and ultimately, to the cost and affordability of food. Several methods have been used for pest and pathogen management including the growing of conventionally bred pest-protected crops, use of chemical pesticides as the primary means of plant protection, and integrated pest management (IPM).

In recent decades, major advances in the science of plant biotechnology have permitted wider access to genetic sources of plant protection against insects and pathogens. Transgenic plants engineered to contain genes for pest-protection have been field tested since 1988 and grown commercially since 1995. From 1995 to 1999, the commercial planting of transgenic pest-protected plants has dramatically increased. Along with these rapid advances in plant biotechnology and its commercial applications, the need to periodically review public oversight and regulation of transgenic plants has emerged.

ES.1 PURPOSE AND SCOPE OF THIS STUDY

In the past, the National Academy of Sciences (NAS) and National Research Council (NRC) have provided guidance to scientists, regulatory agencies, and the public concerning biotechnology and transgenic products. The NRC determined that there was a need for an overview of the current issues surrounding transgenic plants, in particular those engineered to resist pests.[1] As a result, the NRC appointed and funded a committee in 1999 to conduct the study reported here. The committee was charged with the following task:

> The committee will investigate risks and benefits of genetically modified pest-protected (GMPP) plants, and the Coordinated Framework for Regulation of Biotechnology (Coordinated Framework) affecting the use of these plants. The study will 1) review the principles in the NAS Council's white paper, *Introduction of Recombinant DNA-Engineered Organisms into the Environment* (1987), for their continued scientific validity and assess their appropriateness for current decisions regarding GMPP plants, 2) review scientific data which address the risks and benefits of GMPP plants, 3) examine the existing and proposed regulations in light of the identified risks and benefits, 4) examine existing and proposed regulations to qualitatively assess their consequences for research, development, and commercialization of GMPP plants, and 5) provide recommendations to address the identified risks/benefits, and, if warranted, for the existing and proposed regulation of GMPP plants.
>
> Note: The study does not address philosophical and social issues surrounding the use of genetic engineering in agriculture, food labeling, or international trade in genetically modified plants.

As instructed by the charge, the committee focused on transgenic pest-protected plants; however, many of its conclusions and recommendations are applicable to other categories of transgenic plants. Because of public concerns about the safety of our food supply, the committee has placed less emphasis on potential benefits of transgenic pest-protected plants than on potential risks, even when some of these risks seem remote.

During a four-month period, the committee met three times to discuss the issues, review data, and obtain input from the public. Representatives from government-agencies, industry, and nongovernment organizations were invited to discuss the issues and their challenges and concerns. In addition, the committee hosted a public workshop on May 24, 1999, to obtain input from a variety of experts and other interested parties (appendix C). The committee requested data that were submitted for regulatory review of transgenic pest-protected plants from the US

---

[1]For consistency, the committee adopts the broad definition of *pest* used by the statutes which govern the Coordinated Framework for the Regulation of Biotechnology (for example, the Federal Insecticide, Fungicide, and Rodenticide Act and the Federal Plant Pest Act). This definition includes not only invertebrate animals such as insects and nematodes, but also microorganisms such as protozoa, viruses, bacteria, or fungi. In some disciplines, a more narrow definition of pests is used. For example, plant pathologists typically refer to insects as *pests* and disease-causing microorganisms as *pathogens.*

Environmental Protection Agency (EPA), the US Food and Drug Administration (FDA), US Department of Agriculture (USDA), and product registrants (appendix B) and used examples of the data during its analysis.

After reviewing the above information, the committee drafted this report. Chapter 1 introduces the scientific and regulatory issues, chapter 2 focuses on the scientific impacts of conventional and transgenic pest-protected plants, chapter 3 addresses how the scientific information is reviewed in the regulatory framework and presents guiding principles for review, and chapter 4 discusses the positive and negative elements of the current regulatory framework and suggests improvements for the review and exchange of scientific information.

The following pages highlight the committee's major findings, conclusions, and recommendations. Not all of the committee's recommendations could be included in this brief executive summary; therefore, the most general conclusions and recommendations are presented in this section and the more detailed ones are included in chapters 2, 3, and 4.

ES.2   FUTURE STUDIES AND LIMITATIONS
OF THE CURRENT STUDY

This study was conducted with a broad scope and in a short time period in order to provide stakeholders with opportune guidance on a variety of issues. As a result, the committee could not comprehensively analyze all available data on the numerous scientific and regulatory issues. In particular, much data are submitted by developers of transgenic products for regulatory approval (appendix B). The committee could only review examples of such data and of published studies regarding transgenic pest-protected plants.[2] The committee chose examples that covered a range of issues and that were provided by scientific experts representing diverse disciplines and affiliations. The committee focused on the general issues that would be applicable not only to prior product approvals, but also to upcoming decisions related to commercialization.

The committee was able to address several categories of scientific and regulatory issues and develop general conclusions and recommendations to advise researchers, producers, regulators and users of transgenic pest-protected plants. The general conclusions and recommendations identify areas where more analysis is needed. In order to help conduct future analyses, the NRC recently convened a Standing Committee on Biotech-

---

[2]In addition, the committee did not have an opportunity to fully discuss or analyze data published after its last meeting in July 1999. However, some of the more recent information is mentioned in the report.

nology, Food and Fiber Production, and the Environment. This standing committee will identify emerging issues and provide intellectual oversight for subcommittees focusing on particular issues in agricultural biotechnology. Through this mechanism, the NRC expects to publish a series of more detailed, comprehensive reports concerning agricultural biotechnology and looks forward to the opportunity to play a larger role in analyzing and reporting upon the scientific issues.

ES.3   REPORT TERMINOLOGY

ES.3.1   EPA Terminology

The committee recognizes that the term *plant-pesticide*, used by the US Environmental Protection Agency (EPA) to describe the scope of products subject to regulation under its 1994 proposed rule, is controversial. To some extent, the controversy stems from the mistaken impression that EPA will classify plants as pesticides. EPA has consistently stated that the "pesticide" will be defined as the "pesticidal substance that is produced in a living plant and the genetic material necessary for the production of the substance, where the substance is intended for use in the living plant." At least in partial response to the controversy, the agency has recently sought public comment on possible alternatives to the term *plant-pesticide*. The committee agrees that the agency must be sensitive to this issue, but it takes no position on the most appropriate term used for regulatory purposes. Therefore, pesticidal substances, pest protectants, pest resistance genes, and other variations are used throughout this report.

ES.3.2   Genetically Modified Plants

Plant breeders use a variety of genetic techniques to enhance the ability of plants to protect themselves from plant pests. Regardless of the technique used, the committee considers these plants to be genetically modified. Although the committee recognizes that there is no strict dichotomy between the products of conventional and transgenic technologies (see ES.4), in this report it has used the following terms:

> *pest-protected plant or genetically modified pest-protected (GMPP) plant*: refers to any plant that has been genetically modified to express a pesticidal trait[3], regardless of the technique used[4];

---

[3]The committee's definition includes both structural and chemical traits that deter or resist pests.

[4]The committee's definition of pest-protected plants does not include herbicide-tolerant plants.

transgenic pest-protected plant: refers to any plant that has been genetically modified with modern molecular techniques (rDNA technology, commonly referred to as genetic engineering) to express a pesticidal trait;

conventional pest-protected plant: refers to any plant that has been genetically modified by classical or cellular plant breeding techniques (such as hybridization or tissue culture) to express a pesticidal trait.

For completeness, the committee notes that many plants have evolved a natural protection against pests without any type of genetic modification done by humans. This report refers to those plants as *naturally pest-protected plants*.

ES.4 REVIEW OF THE
1987 NATIONAL ACADEMY OF SCIENCES PRINCIPLES

As the first assigned task, the committee reviewed the 1987 NAS white paper, *Introduction of Recombinant DNA-Engineered Organisms into the Environment: Key Issues*. The 1987 paper focused on the safety of rDNA techniques and on ecological issues associated with the potential spread of transgenic organisms or genes associated with transgenic organisms, and it provided the following conclusions:

- point 1 "There is no evidence that unique hazards exist either in the use of rDNA techniques or in the movement of genes between unrelated organisms."
- point 2 "The risks associated with the introduction of rDNA-engineered organisms are the same in kind as those associated with the introduction of unmodified organisms and organisms modified by other methods."
- point 3 "Assessment of the risks of introducing rDNA-engineered organisms into the environment should be based on the nature of the organism and the environment into which it is introduced, not on the method by which it was produced."

The committee discussed the above principles in light of its knowledge of the underlying scientific processes involved in conventional and transgenic methods. It is important to point out that the committee is not aware of controlled field studies which directly compare the ecological effects of transgenic and conventional pest-protected plants bred for the same pesticidal traits. Therefore, the committee's conclusions about the 1987 NAS principles are not based on data from such comparisons, but on mechanistic knowledge and scientific information about the resulting genetically modified plants. For example, conventional breeding often involves the transfer of traits which are controlled by several interacting genes and often occurs without specific knowledge of which genes and gene products are involved. Therefore, some of the plants produced by this method could have unanticipated properties. With transgenic methods, there is often more knowledge about the genes and gene products being transferred, but diverse traits and genes from unrelated organisms can be transferred so some specific products could have unique properties. Because both methods have the potential to produce organisms of high or low risk, the committee agrees that the *properties* of a genetically modified organism should be the focus of risk assessments, not the *process* by which it was produced (point 3).

The committee also agrees with points 1 and 2 in the sense that the potential hazards and risks associated with the organisms produced by conventional and transgenic methods fall into the same general categories. As this report discusses, toxicity, allergenicity, effects of gene flow, development of resistant pests, and effects on non-target species are concerns for both conventional and transgenic pest-protected plants. In this regard, the committee found no strict dichotomy between, or new categories of, the health and environmental risks that might be posed by transgenic and conventional pest-protected plants (points 1 and 2), and recognizes that the magnitude of risk varies on a product by product basis (point 3).

The present committee found the three general principles to be valid within the scope of issues considered by the 1987 paper, and the present report further clarifies and expands on these principles.

This report expands on the 1987 principles by describing various methods of both conventional and transgenic plant breeding, and their potential consequences.

ES.5   POTENTIAL HEALTH AND ECOLOGICAL IMPACTS AND RESEARCH NEEDS

Conventional pest-protected plants have substantially improved plant health and agricultural productivity and have often lessened the need for chemical pesticides. Transgenic pest-protected plants have the potential to make similar contributions, as has already been documented with transgenic pest-protected cotton (section 1.5.5). Human health and environmental benefits could arise from reductions in the application of chemical pesticides resulting from the commercial production of certain transgenic pest-protected plants. However, the relative risks and benefits will depend on the particular transgenic pest-protected plant in question.

Historically, pest-protected plants have rarely caused obvious health or environmental problems, but there is a potential for undesirable effects. Therefore, a major goal for further research and development of transgenic and conventional pest-protected plants should be to enhance agricultural productivity in ways that also foster more sustainable agricultural practices, enhance the preservation of biodiversity, and decrease the potential for health problems that could be associated with some types of pest-protected plants. Although the committee focused its discussions on transgenic pest-protected plants, many of the following recommendations for research and development also apply to conventional pest-protected plants.

ES.5.1   Health Impacts And Research Needs

Health impacts that the committee considered fall into three general categories: allergenicity, toxicity, and pleiotropic[5] effects of genetic modifications.

The potential for allergenic responses to novel gene products was considered. Such responses have not been documented for commercialized transgenic pest-protected plants, although one incident has been documented at the research stage. Several indirect tests for allergenicity are available. For novel proteins, the most common methods involve analyzing the protein for its digestibility, estimating the level of protein expression and consumption, and assessing homology to known allergens. While these indirect tests can be good indicators of potential allergenicity, the development of more direct tests is highly desirable. Therefore, the committee recommends that

Priority should be given to the development of improved methods for identifying potential allergens in pest-protected plants, specifically, the development of tests with human immune-system endpoints and of more reliable animal models.

The committee reviewed data concerning toxicity testing and potential pleiotropic or secondary effects of genetic modification. The committee concluded that monitoring for pleiotropic changes in plant physiology and biochemistry during the development of pest-protected plants should be an important element of health-safety reviews, in addition to testing the toxicity of the introduced gene products (see ES.6.4). Although results of tests for changes in the levels of certain endogenous plant toxi-

---

[5]Defined as *simultaneous effects on more than one character of the organism.*

cants are presented during consultation with FDA, there is a lack of an extensive database on the natural levels of such compounds in both transgenic and conventional pest-protected plants. The committee recognizes the challenges associated with detecting changes in those compounds given insufficient analytical information, and therefore, recommends research to

Assess and enhance data on the baseline concentrations of plant compounds of potential dietary or other toxicological concern, and determine how concentrations of these compounds may vary depending on the genetic background of the plant and environmental conditions.

In addition to the above research, the committee recommends that

The EPA, FDA, and USDA collaborate on the establishment of a database for natural plant compounds of potential dietary or other toxicological concern.

The committee recognizes that a significant amount of time and resources will be needed to establish such a database, given the complexity of these plant compounds.

For some novel pest-protectants developed for future commercialization, longterm toxicity testing may be warranted. Tests which involve feeding of large quantities of pest-protected plants to animals have limitations, and the results can be difficult to interpret especially when the animal's natural diet does not consist of the type and quantities of the plant being tested (section 2.5.2). Therefore, the committee recommends research to

Examine whether longterm feeding of transgenic pest-protected plants to animals whose natural diets consist of the quantities and type of plant material being tested (for example, grain or forage crops fed to livestock) could be a useful method for assessing potential human health impacts.

In conclusion, although there is the potential for the adverse health effects discussed in this section,

The committee is not aware of any evidence that foods on the market are unsafe to eat as a result of genetic modification.

ES.5.2 Ecological Impacts and Research Needs

Three major ecological impacts[6] were considered by the committee: effects on nontarget[7] species, effects of gene flow[8], and evolution of pest resistance to pest-protected plants.

The committee reviewed studies concerning nontarget effects. The committee found that both conventional and transgenic pest-protected crops could have effects on nontarget species, but these potential effects are generally expected to be smaller than the effects of broad-spectrum synthetic insecticides. Therefore, the use of pest-protected crops could lead to greater biodiversity in agroecosystems where they replace the use of those insecticides (section 2.6.3). The use of transgenic pest-protected plants should also be compared with sustainable agriculture methods for crop protection. The committee recommends research to

Determine the impacts of specific pest-protected crops on nontarget organisms, compared with impacts of standard and alternative agricultural practices through rigorous field evaluations.

Gene flow between cultivated crops and wild relatives was the second ecological impact considered by the committee. On the basis of the literature, the committee found that pollen dispersal can lead to gene flow among cultivated crops and from cultivated crops to wild relatives but that only trace amounts of pollen are typically dispersed further than a few hundred feet (section 2.7). The committee found that the transfer of either conventionally bred or transgenic resistance traits to weedy relatives potentially could exacerbate weed[9] problems, but such problems have not been observed or adequately studied. Therefore, the committee recommends further research to

Assess gene flow and its potential consequences: develop a list of plants with wild or weedy relatives in the United States; identify key factors that regulate weed populations; assess rates at which pest resistance genes from the crop would be likely to spread among weed populations; and evaluate the impact of specific, novel resistance traits on the weed abundance.

---

[6]The committee's ecological assessment focused on potential impacts of food and fiber crops, not on the potential impacts of other types of transgenic pest-protected plants that might be commercialized in the future (for example, forest trees).

[7]Organisms that are not the target for the particular plant-pesticide.

[8]The transfer of genetic information from one organism to another.

[9]The committee's definition of a weed includes plants that are unwanted in human-dominated or natural habitats.

Develop transgenic or other techniques that decrease potential for the spread of transgenes into wild populations.

Evolution of pest resistance to pest-protected plants was the third major ecological impact addressed by the committee. The committee concluded that pest resistance to pest-protected plants could have a number of potential environmental and health impacts such as a return to the use of more harmful chemicals or replacement of an existing pest-protected variety with novel varieties for which there is less information available about health and environmental impacts. The committee recommends that

If a pest-protectant or its functional equivalent is providing effective pest control, and if growing a new transgenic pest-protected plant variety threatens the utility of existing uses of the pest-protectant or its functional equivalent, implementation of resistance management practices for all uses should be encouraged (for example, Bt proteins used both in microbial sprays and in transgenic pest-protected plants).

In addition to the above recommendations, the committee recommends general ecological research to

Improve our understanding of the molecular basis of pest-plant interactions and of the population ecology and genetics of target pests so that more ecologically and evolutionarily sustainable approaches to the use of pest-protected plants can be developed.

Develop more specific expression systems for transgenes in ways that lessen nontarget exposure and delay pest adaptation (for example, use of promoters[10] that would limit expression to certain tissues).

Monitor ecological impacts of pest-protected crops on a long term basis to ensure the detection of impacts that may not be predicted from tests conducted during the regulatory approval process.

ES.6   THE COORDINATED FRAMEWORK FOR REGULATION

ES.6.1   Background and History

In 1986, the Coordinated Framework for the Regulation of Biotechnology apportioned jurisdiction over transgenic products by using exist-

---

[10]DNA sequences which regulate the expression of genes.

ing legislation: for example, plants came under the jurisdiction of the Federal Plant Pest Act (FPPA) administered by the USDA; food and feed under the jurisdiction of the Federal Food, Drug, and Cosmetic Act (FFDCA) administered by the FDA; and microorganisms and substances used for pest control under the jurisdiction of the Federal Insecticide, Fungicide, and Rodenticide Act (FIFRA) and parts of FFDCA, administered by the EPA. Transgenic pest-protected plants were not addressed in the original framework document.

USDA published its policy under the coordinated framework providing for field testing permits for transgenic plants in 1987 and field testing notifications in 1993 and 1995. In 1993, it finalized its policy for determining when certain plants would no longer be regulated articles. In 1992, FDA published its policy for foods derived from new plant varieties based on its role under FFDCA. In 1994, EPA proposed a rule to regulate the pesticidal substances in pest-protected plants as plant-pesticides under FIFRA and FFDCA. Several groups opposed that statutory interpretation on both legal and scientific grounds; others supported the EPA's oversight of transgenic pest-protected plants, given the agency's mission to address environmental concerns. In the last few years, there have been concerns expressed by several professional societies and other groups over the broad scope of the proposed EPA rule and opposite concerns expressed by consumer and environmental groups that the EPA rule does not adequately cover all of the risk issues.

ES.6.2   Overall Approach

The committee recognizes that

There is an urgency to complete the regulatory framework for transgenic pest-protected plant products because of the potential diversity of novel traits that could be introduced by transgenic methods and because of the rapid rate of adoption of and public controversy regarding transgenic crops.

Accordingly, the committee has chosen to take EPA's proposed rule and the overarching coordinated framework as given and as designed for transgenic products[11], and to examine ways in which this current regulatory approach and its use of scientific information might be improved. In so doing, the committee does not suggest that this is the only possible approach to regulating these products. It is beyond this committee's

---

[11]Although the committee focuses on the regulation of transgenic pest-protected plants, conventional pest-protected plants are discussed for scientific comparisons.

scope to determine which of the three federal agencies (USDA, EPA, or FDA) is best suited to regulate pesticidal substances expressed in transgenic plants.

EPA's current proposal for regulating pesticidal substances in pest-protected plants claims broad jurisdiction over such products in all seeds and plants sold with claims of pest-protection, but it grants a generic exemption from registration to those bred by conventional means. The committee agrees with EPA's proposed exemption of pesticidal substances in conventionally bred plants, because the committee recognizes that there are practical reasons for exempting those substances based in part on historical experience of safe use of, and the benefits provided by these crops. However, the committee questions the *scientific* basis used by EPA for this exemption because there appears to be no strict dichotomy between the risks to health and the environment that might be posed by conventional and transgenic pest-protected plants.

The committee found that, in some cases, the use of conventional pest-protected crops might have the potential to lead to human and animal health impacts; therefore

There is a need to significantly increase research aimed at assessing the potential risks posed by conventional pest-protected plants, and make improvements of conventional breeding procedures, if found appropriate.

ES.6.3   Scientific Basis for the 1994 Proposed EPA Rule

Consistent with the coordinated framework and its statutory mandates, EPA has asserted jurisdiction over pesticidal substances in transgenic pest-protected plants in its 1994 proposed rule. The committee reviewed the scientific basis of EPA's 1994 proposed rule and the exemption of certain categories of *transgenic* pest-protected plants under this rule. The committee found most of the criteria used by EPA for assessing *transgenic* pest-protected products to be scientifically valid, but there were some exceptions.

EPA proposes to exempt all plant-pesticides where the structural gene for producing the plant-pesticide is derived from a sexually compatible plant. The committee found that the current EPA rule would exempt transgenic pest-protectants if the structural gene came from a sexually compatible plant, regardless of the source of the promoter for expression of the gene. This categorical exemption of transgenic pest-protectants derived from transgenes from sexually compatible plants could result in no EPA regulation of genetically engineered products which contain higher levels of toxicants. The committee agrees that, in many cases, exemptions for certain sexually-compatible transgenic pest-protectants will be warranted; however, it questions the *categorical* exemption of these products. The committee recommends that

Given that transfer and manipulation of genes between sexually compatible plants could potentially result in adverse effects in some cases (for example, modulation of a pathway that increases the concentration of a toxicant), and given the public controversy regarding transgenic products, EPA should reconsider its *categorical* exemption of *transgenic* pest-protectants derived from sexually compatible plants.

The committee also examined EPA's proposed exemption for viral coat proteins[12] expressed in transgenic pest-protected plants. Viral coat proteins in transgenic pest-protected plants are not expected to jeopardize human health, inasmuch as consumers already ingest these substances in nontransgenic food, so the committee agrees with the exemption of these proteins from EPA jurisdiction under FFDCA. However, the committee questions the EPA's *categorical* exemption of all viral coat proteins under FIFRA due to concerns about the potential for outcrossing with weedy relatives. The committee agrees that exemption of particular viral coat proteins in certain plant species will be warranted. However, the committee suggests that

EPA should not *categorically* exempt viral coat proteins from regulation under FIFRA.

ES.6.4 Scientific Data Used by the Agencies in the Regulatory Process

The committee reviewed examples of data submitted by applicants to the regulatory agencies for currently commercialized transgenic pest-protected plant products (that is, products with Bt and viral coat proteins). The federal agencies already address most of the categories of scientific concerns presented in this report (see table 4.3). However, the committee found some areas where the risk assessment process for transgenic pest-protected plants could be improved.

In reviewing toxicity testing relevant to human health, the committee found that,

When the active ingredient of a transgenic pest-protected plant is a protein and when health effects data are required, both short-term oral

---

[12]Virus-derived proteins that form a capsule around viral DNA or RNA.

toxicity and potential for allergenicity should be tested.   Additional categories of health effects testing (such as for carcinogenicity) should not be required unless justified.

Additional categories of toxicity testing do not appear justified for currently commercialized products such as many Bt proteins (Cry1A and Cry3A) and viral coat proteins. However, it is important that the tests that are performed be rigorous, logical, and scientifically sound. Novel or less familiar plant-pesticides (that is, in comparison to viral coat proteins and Bt toxins) may require additional categories of toxicity testing.

Although the committee realizes that it is often difficult to obtain enough plant-expressed protein for toxicological testing; tests should be conducted whenever possible using the protein as it is expressed in the plant. The committee recommends that

The EPA should provide clear, scientifically justifiable criteria for establishing biochemical and functional equivalency when registrants request permission to test non plant-expressed proteins in lieu of plant-expressed proteins.

In addition to human health toxicity testing, allergenicity testing is very important. The committee recognizes that the FDA has developed preliminary information on the assessment of potential food allergens that could be helpful to applicants as they evaluate potential products and develop product-specific data to address questions concerning allergenicity. The committee recommends that

FDA should put a high priority on finalizing and releasing preliminary guidance on the assessment of potential food allergens, while cautioning that further research is needed in this area.

The committee found some room for improvement in the procedures used in USDA's review of outcrossing or gene flow for virus-resistant squash (section 3.1.4). USDA's commercialization of the squash was controversial because the transgenic squash potentially could transfer its acquired virus-resistance genes via pollination to wild squash (*Cucurbita pepo*), which is an agricultural weed in some parts of the southern United States. USDA's assumption that transgenic resistance to viruses will not affect the weediness of wild relatives might be correct, but longer-term empirical studies are needed to determine whether this is true. The committee recommends that

USDA should require original data to support agency decision-making concerning transgenic crops when published data are insufficient.

ES.7 OPERATIONAL ASPECTS AND IMPACTS OF THE COORDINATED FRAMEWORK

ES.7.1 Elements of an Effective Regulatory Framework

The committee finds that, operating under the coordinated framework, EPA, USDA, and FDA have successfully applied existing statutes to address the introduction of transgenic pest-protected plant products, but concludes that there is room for improvement. In particular, those agencies have achieved a significant degree of coordination in their oversight of transgenic pest-protected plants, but certain aspects of this coordination could be enhanced. Only through effective coordination can the three lead agencies minimize duplication, avoid inconsistent regulatory decisions, address potential gaps in oversight, and ensure that regulations evolve with experience and scientific advancements. Ultimately, the credibility of the regulatory process and acceptance of products of biotechnology depend heavily on the public's ability to understand the process and the key scientific principles on which it is based.

The committee identified five elements of an effective regulatory system which support the objectives of the coordinated framework (Box ES.1).

For example, to improve the transparency of the regulatory process under the coordinated framework, the committee recommends that

The quantity, quality and public accessibility of information on the regulation of transgenic pest-protected plant products should be expanded.

The USDA-sponsored coordinated framework database to link agencies' regulations and decisions (USDA 1999e) is useful, but should be

---

Box ES.1
Elements that Support the Objectives of the
Coordinated Framework

- Consistency of definitions and regulatory scope.

- Clear establishment of lead and supporting agencies with a mechanism for effective interagency communication.

- Consistency of statements of information to support reviews.

- Comparably rigorous reviews.

- Transparency of review process.

expanded by all three agencies to include more public information about specific products and to link agencies' decisions about specific products. The EPA pesticide fact sheets for transgenic plant pesticides should be improved because they currently do not clearly and quantitatively present the results of safety testing.

Another element in box ES.1 is consistency of regulatory scope. The scope of agency oversight, in some cases, needs to be clarified (see section 4.3.3).

With new recombinant DNA methods, USDA can no longer rely on the production of transgenic pest-protected plants with regulatory sequences[13] from plant pests (for example, *Agrobacterium tumefaciens* vectors and cauliflower mosaic-virus promoters). Some new products may be developed using natural plant regulatory sequences. It is not clear if USDA would consider these products "plant pests." Therefore, the committee recommends that

The USDA should clarify the scope of its coverage as there are some transgenic pest-protected plants that do not automatically meet its current definition of a plant pest.

The delineation of lead and supporting agency jurisdiction over transgenic pest-protected plant products is generally well defined. Agency reviews generally lack duplication and achieve consistency. However, the committee identified some examples where communication and coordination could be improved.

To improve coordination among the three regulatory agencies, EPA, FDA, and USDA should develop a memorandum of understanding (MOU) for transgenic pest-protected plants that provides guidance to identify the regulatory issues that are the purview of each agency (for example, ecological risk and pesticide tolerance assessment for EPA, plant pest risk for USDA, and dietary safety of whole foods for FDA), identifies the regulatory issues for which more than one agency has responsibility (for example, gene flow for EPA and USDA and food allergens for EPA and FDA), and establishes a process to ensure appropriate and timely exchange of information between agencies.

If differences in regulatory findings remain after interagency consultations, they should be adequately explained to ensure that regulatory decisions are not in conflict and do not have the appearance of conflict.

------

[13]Non-coding regions of genes which are involved in controlling the expression of genes.

The committee found that the three agencies have common data requirements specifically for biology of the recipient plant, molecular biology methods used to develop the product, identification and characterization of inserted genetic material and its product(s), and identity and characterization of selectable markers. Therefore, the committee recommends that

To enhance consistency of review, EPA, USDA, and FDA should develop a joint guidance document for applicants that identifies the common data and information the three agencies need to characterize products.

Taking into account the above suggestions, the committee hopes that the regulatory framework for transgenic pest-protected plants can be quickly completed by clarifying, revising, and finalizing the EPA 1994 proposed rule; publishing guidance on regulatory requirements; and developing additional interagency MOUs. However, once established, the committee recommends that

Regulations should be considered flexible and open to revision, so that agencies can adapt readily to new information and improved understanding of the science that underlies regulatory decisions. The agencies have attempted to maintain a dynamic regulatory process, but more could be done to retain flexibility in the future (see chapter 4).

ES.7.2 Economic Costs Associated With Regulation

Positive impacts of regulation might include reduced health and environmental effects and increased consumer confidence in the food supply. However, there are also economic costs associated with the regulation of transgenic pest-protected plants. The committee reviewed an analysis on the economic costs of regulation (section 4.4 and appendix A[14]). From this review and other discussions in chapter 4 (see sections 4.2 and 4.3), the committee concludes that regulators should be sensitive to the unique issues facing researchers, plant breeders, and seed distributors, particularly those in the public sector or those who have not traditionally been subject to federal regulation. In particular, the committee recommends that

---

[14]This appendix was authored by an individual committee member and is not part of the committee's consensus report. The committee as a whole may not necessarily agree with all of the contents of appendix A.

Regulatory agencies should aggressively seek to reduce regulatory costs for small biotechnology startup companies, small to medium size seed companies, and public sector breeders by providing flexibility with respect to data requirements, considering fee waivers wherever possible, and helping these parties navigate their regulatory systems.

The committee does not recommend waiving necessary regulatory requirements; however, where regulation is not warranted, agencies should look for appropriate opportunities to promote nonregulatory mechanisms to address issues associated with transgenic pest-protected plant products, including encouraging development of voluntary industry consensus standards and product stewardship programs.

ES.8   STRIVING FOR THE IDEAL REGULATORY FRAMEWORK

In the time allotted for this report, the committee focused on providing meaningful input to improve the review of scientific data under the coordinated framework and the proposed EPA plant-pesticide rule. The committee's findings, conclusions, and recommendations will need to be tested before they are confirmed as useful methods to enhance scientific review during the regulation of transgenic pest-protected plants. The committee realizes that these improvements may not be possible without increased resources for the federal agencies involved in agricultural biotechnology and for research focused on the risks and benefits. A solid regulatory system and scientific base are important for acceptance and safe adoption of agricultural biotechnology, as well as for protecting the environment and public health. In general, the current US coordinated framework has been operating effectively for over a decade. However, the committee has identified several kinds of improvements that would be helpful in the face of a larger number of commercialized transgenic pest-protected plants and novel gene products introduced into these plants. Those improvements might be necessary for increased confidence in US agricultural biotechnology both domestically and worldwide.

1

Introduction and Background

1.1 THE IMPORTANCE OF CROP PROTECTION

Farmers have been trying to minimize the impacts of crop pests for thousands of years. Insects, nematodes, bacteria, fungi, and viruses can cause massive destruction of important crops, and this destruction can have great socioeconomic effects. For example, the Irish potato famine of the 1800s led to the deaths of about 1 million people and large-scale emigration. More recently, head blight caused by the fungal pathogens *Fusarium graminearum* and *F. poae* caused about $3 billion in damage to wheat and barley in 1991-1996 (US Wheat and Barley Scab Initiative 1998). An estimated $7 billion in crop losses per year in the United States are caused by nematodes (NSTC 1995) and even greater losses are caused by arthropod pests. In addition to socioeconomic effects, some plant pests pose human health hazards, such as those caused by fungal mycotoxins. Pest control is a continuous process: as pest-protected plants are bred or new chemical pesticides are developed, pests evolve to overcome these control methods.

Early methods to control pests include the use of sulfur fumigation in 1000 BC, ants for biocontrol in 324 BC, and crop rotation, controlled irrigation, and manure application during the Roman Empire. Arsenic was used in the 1600s, *Bacillus thuringiensis* was developed as a microbial insecticide as early as 1938 (NRC 1996), and the use of synthetic pesticides became the predominant means of pest control in the 1940s. Since the 1960s, there has been wide implementation of integrated pest-management (IPM) approaches, designed to use a variety of natural controls and

19

cultural methods to suppress pest populations (Smith and Van denBosch 1967). IPM is an approach which manages pests by biologically integrated alternatives for pest control (US Congress 1947, as amended in the 1972 Federal Environmental Pesticide Control Act, section 136r(a)) and is "a sustainable approach to managing pests by combining biological, cultural, physical, and chemical tools in a way that minimizes economic, health, and environmental risks" (US Congress 1947, as amended by the 1996 Food Quality Protection Act, section 136r-1). Pesticides are used only as necessary and when other control methods have failed (Stern et al. 1959).

1.2   DIVERSE GENETIC MODIFICATION METHODS

To develop pest-resistant or tolerant cultivars, plant breeders have taken advantage of natural genetic variation or induced mutations. The methods that plant breeders use depend on the type of cultivar they want to improve (for example, an inbred line, a hybrid, or a population) and the reproductive biology of the plant (for example, self-pollinated or cross-pollinated) (Fehr 1987; Stoskopf et al. 1993).

An inbred line (or purebred) is phenotypically uniform[1], and the progeny[2] are identical with the parent. Many self-pollinated crops are released as inbred lines (for example, soybeans, *Glycine max*, and barley, *Hordeum vulgare*). A hybrid is the cross between two or more inbred lines; it can also be phenotypically uniform but not genetically identical with the parents. Many cross-pollinated crops are released as hybrids (for example, corn or maize, *Zea mays*). A plant population results from crossing a number of lines and is genetically and phenotypically diverse, although for key traits, a population can be phenotypically uniform (for example, every plant resistant to a pest).

All genetic modification methods for crop improvement consist of introducing variation, selecting useful variants, and field-testing the selected lines, hybrids, or populations to determine their merit. In the past, almost all commonly used plant breeding techniques began with artificial crosses, in which pollen from one plant is transferred to a reproductive organ of another, sexually compatible plant. Crossing allows for the combining of desirable traits, such as pest resistance and increased yield, from two or more plant cultivars.[3] The objective is to combine these traits in a new cultivar that is superior to its parents. To overcome some of the

---

[1]Expressing the same phenotypes or traits.
[2]Offspring.
[3]Cultivated variety of plant.

barriers to sexual hybridization between cultivated and wild relatives, rescue of pollinated embryos has been used: when a cross yields a viable embryo but the surrounding seed endosperm[4] is not viable, the embryo is taken from the nonviable seed environment and "rescued" by being grown in tissue culture.

Other techniques to introduce variation in cultivars include cell fusion, somaclonal variation, chemical or x-ray mutagenesis, and genetic engineering (see section 2.4.2). Cell fusion is used to produce novel combinations of genomic material from nuclei and organelles when plants are not sexually compatible (Ehlenfeldt and Helgeson 1987); it can be performed only on plants that can be cultured with protoplast technologies. With protoplast technologies, cells are disconnected from tissues, their walls are removed, and their membranes are prepared for fusion. Somaclonal variation is variation that occurs during the tissue-culture process, and its phenotypic outcomes are often similar to other forms of mutagenesis. Genetic engineering is the transfer of a or a few genes into a cultivar with the use of *Agrobacterium tumefaciens,* microprojectile bombardment, electroporation, or microinjection. Transgenic methods will be discussed in more detail in subsequent sections of this report.

One of the main differences among the techniques used for introducing variation is in the amount of DNA involved. In progeny resulting from a cross between two cultivars, half the genome comes from each cultivar. Each half (haploid) genome contains a significant amount of DNA (table 1.1). The amount transferred with conventional breeding in the case of *Arabidopsis* could be 70 megabases (Mb) (half the progeny's haploid genome comes from each parent). For bread wheat, the amount of DNA could be almost 8000 Mb. In contrast, transgenic methods involve the addition of only a few genes and flanking regulatory sequences (totaling about 1-20 kilobases).

Another important difference among the techniques can be the source of the transferred DNA. Sexual hybridization involves genes from sexually compatible species, which tend to be rather similar. Mutagenesis and the somaclonal variation process do not add genes, but rather modify existing genes. Cell fusion can add genes from evolutionarily divergent plant species (such as, plants from different genera), but normally fused cells are from somewhat related plants (for example, the technique has not been conducted by fusing cells from plants and microorganisms). In genetic engineering or transgenic methods, genes from any organism in the biosphere can be used as long as the regulatory sequences are functional in the host plant. For example with genetic engineering researchers

---

[4]Nutritional tissue in seeds.

TABLE 1.1   Genome Size of Common Plants

Common Name	Scientific Name	Nuclear DNA Content	
		Picograms (diploid nucleus)[a]	~Millions of base pairs (haploid nucleus)[b]
Arabidopsis	*Arabidopsis thaliana*	0.3	145
Barley	*Hordeum vulgare*	10.1	4873
Brussels sprout	*Brassica oleracea* ssp. *gemmifera*	1.3	628
Corn	*Zea mays*	4.75-5.63	2292-2716
Cotton	*Gossypium hirsutum*	4.39	2118
Oats	*Avena sativa*	23.45	11315
Papaya	*Carica papaya*	0.77	372
Peanut	*Arachis hypogeae*	5.83	2813
Rice	*Oryza sativa*	0.87-0.96	419-463
Soybean	*Glycine max*	2.31	1115
Tobacco	*Nicotiana tabacum*	8.75-9.63	4221-4646
Tomato	*Lycopersicum esculentum*	1.88-2.07	907-1000
Bread Wheat	*Triticum aestivum*	33.09	15966
Wild Wheat	*Triticum monococcum*	11.92	5751

[a]1 picogram = 965 million base pairs, haploid nucleus

[b]DNA content of unreplicated haploid chromosome complement

Source: Data from Arumuganathan and Earle, 1991.

have added genes to potatoes from bacteria, viruses, chickens, and moths. The foreign gene can also be modified by molecular techniques before introduction into the plant (for example, by incorporating DNA base pair substitutions).

However, a key question is whether the fact that genes can be obtained from broader sources for plant biotechnology inherently impacts the safety of the resulting genetically engineered organism (see sections 2.2.1 and 2.4.2). Foreign genes engineered into plants may or may not be homologous to genes already present in the plant or the food supply.

1.3   HISTORY AND IMPACT OF BREEDING METHODS

Selection for desirable traits and hybridization has been used since the advent of human agriculture, but the logic underlying the inheritance of traits was not discovered until the middle 1800s. In the 1860s, Gregor Mendel demonstrated the process of heredity by hybridizing different varieties of pea (*Pisum sativum*) and examining traits such as flower and seed color, seed and pod shape, flower position, and plant height in subsequent generations. His revolutionary experiments paved the way for modern agriculture by showing that through controlled pollination crosses, characteristics are inherited in a logical and predictable manner. In 1905, Roland Biffen, of England, built on Mendel's experiments by illustrating that the ability of wheat (*Triticum aestivum*) to resist a rust fungus could be passed to later generations (NAS 1998).

Since then, many plants have been bred to include desirable traits, such as pest resistance. Blight resistance traits from a Mexican potato species (*Solanum demissum*) have been introduced into over 50% of all potato cultivars (NRC 1989). Blight-resistant corn (*Zea mays*), rust-resistant wheat (*T. aestivum*), and aphid-resistant alfalfa (*Medicago sativa*) are other notable examples of conventional plant breeding. Major gains in crop yields have been attributed partially to advances in classical plant breeding and plants developed for pest resistance. Corn yields have increased from 5 metric tons per hectare in 1967 to 8 metric tons per hectare in 1997, cereal harvests have been increasing at an average rate of 1.3% per year (Mann 1999), world food production has doubled since 1960, and agricultural productivity from land and water use has tripled (NSTC 1995).

Conventional breeding will likely continue to play an essential role in the improvement of agricultural crops. However, many believe that traditional breeding methods will not be sufficient to meet increasing demands in developing countries for staple crops, such as wheat (*T. aestivum*), rice (*Oryza sativa*) and corn (*Zea mays*) (Mann 1999). Classical methods are time-consuming (that is they take approximately 10 years to develop a variety) and labor-intensive (only one line of thousands becomes a useful variety). In addition, beneficial traits can be linked to or lead to undesirable traits, such as disease susceptibility. For example, when male-sterile corn was extensively grown in the 1960-1970s to promote hybrid-corn production, a new race of southern corn leaf blight fungus (*Helminthosporium maydis*) evolved which successfully attacked this type of corn and significantly decreased US corn yields (Dewey et al. 1988). Some have proposed transgenic methods to augment the advances in conventional breeding.

1.4 EMERGENCE OF RECOMBINANT DNA AND OVERVIEW OF TRANSGENIC PEST-PROTECTED PLANTS

In the past two decades, scientists have focused on expanding genetic modification methods to include the use of recombinant DNA (rDNA) techniques. New varieties generally can be produced faster by rDNA than by conventional breeding methods. rDNA methods allow the introduction of genes from distantly related species or even from different biologic kingdoms. In addition, detailed knowledge of the trait being introduced (such as a DNA sequence or cellular function) can lead to less variability in the offspring and eliminate some of the uncertainty about linked traits. The site of insertion of a gene can affect its expression, but generally plants with the appropriate level of expression can be selected if a number of transgenic plants are produced. After a trait is introduced by transgenic methods, the resulting plant can be sexually hybridized with useful varieties developed by conventional breeding.

1.4.1 Emergence of Recombinant DNA Methods

Recombinant DNA methods emerged in the early 1970s after the discovery of restriction enzymes (Linn and Arber 1968; Meselson and Yuan 1968), DNA-sequencing methods (Sanger and Coulson 1975; Maxam and Gilbert 1977), and plasmid and viral vectors for engineering organisms (Jackson et al. 1972; Cohen et al. 1973). The methods have been used ever since to manipulate DNA fragments that contain genes of interest.

With the advent of this technology, concerns about the safety of experiments that use rDNA methods developed. The National Academy of Sciences (NAS) in 1974 convened a committee to assess the safety concerns associated with rDNA research. The committee recommended that rDNA experiments be postponed until further evaluation of the risks (Berg et al. 1974). Soon after, the International Conference on Recombinant DNA Molecules, better known as the Asilomar Conference, was held (Berg et al. 1975). An outline of guiding principles and restrictions for rDNA research was generated at this conference. In 1976, the principles were reviewed by the National Institutes of Health (NIH), which implemented official guidelines to be administered by the NIH Recombinant DNA Advisory Committee (RAC) (NIH 1976). The guidelines focused on laboratory containment of rDNA microorganisms. As more experiments were conducted and more data on the risks were generated, less restrictive guidelines were put into place (NIH 1978). In recent years, the guidelines have been expanded to include other rDNA applications, such as gene therapy, and have been adopted not only by institutions receiving federal funding, but also by industry and state institutions.

About a decade after the emergence of rDNA technology, genes from the bacterium *Agrobacterium tumefaciens* were used to carry foreign genes into plants. *Agrobacterium* inserts portions of its tumor-inducing (T_i) plasmid-encoded genes into plant chromosomes as part of its natural, parasitic life cycle (Nester et al. 1983). When researchers add foreign DNA (such as a gene for pest-protection) in between T_i plasmid-encoded insertion sequences, the foreign DNA sequences are also inserted into the plant's chromosome. The first transgenic plants were developed with

Agrobacterium-transformation methods (Horsch et al. 1985). Since then, other methods for plant transformation, such as electroporation and particle-gun transformation, have been developed (Klein et al. 1987; Finer et al. 1999); these methods allow transformation of plants that are not natural hosts for *Agrobacterium*.

1.4.2 Development of a Regulatory Framework for Transgenic Plants

Concurrently with developments in the technical aspects of genetically engineering crops by using rDNA methods, regulatory concerns about the release of genetically engineered organisms into the environment emerged. The NIH guidelines in 1978 prohibited the environmental release of genetically engineered organisms unless exempted by the NIH director. In 1982, the RAC reviewed a request to field test "ice-minus" bacteria, strains of *Pseudomonas syringae* and *Erwinia herbicola* that had inactivated ice-nucleation genes (Lindow and Panopoulos 1988). NIH approved the request in 1983 (NIH 1983). The approval of the field trial was controversial and sparked several court cases that invoked the National Environmental Policy Act (NEPA) (US Congress 1969). NEPA requires that any agency decision that significantly affects the quality of the environment be accompanied by a detailed statement or an assessment of the environmental impacts of the proposed action and of alternatives to it.

As the field trial was being debated by the courts, a congressional hearing was held at which questions were raised about the ability of federal agencies to address hazards to ecosystems in light of the uncertainties (US Congress 1983). At a second hearing in 1984, the Senate Committee on Environment and Public Works discussed the potential risks with representatives of the Environmental Protection Agency (EPA), NIH, and the US Department of Agriculture (USDA). The government agencies stated that existing statutes were sufficient to address the environmental effects of genetically engineered organisms (US Senate 1984). Also in 1984, a White House committee was formed under the auspices of the Office of Science and Technology Policy (OSTP) to propose a plan for regulating biotechnology.

In 1986, OSTP published the Coordinated Framework for the Regulation of Biotechnology (OSTP 1986), which is still used today. The framework is based on the principle that techniques of biotechnology are not inherently risky and that biotechnology should not be regulated as a process, but rather that the products of biotechnology should be regulated in the same way as products of other technologies. The coordinated framework outlined the roles and policies of the federal agencies and contained the following ideas: existing laws were, for the most part, adequate for oversight of biotechnology products; the products, not the process, would be regulated; and genetically engineered organisms are not fundamentally different from nonmodified organisms. A 1987 National Academy of Sciences white paper came to similar conclusions, recommending regulation of the product, not the process, and stating that genetically engineered organisms posed no new kinds of risks, that the risks were "the same in kind" as those presented by nongenetically engineered organisms (NAS 1987) (section 2.2.1).

The coordinated framework considered existing regulations and laws that were potentially applicable to biotechnology and proposed how EPA, USDA, and the Food and Drug Administration (FDA) would cooperate to review the safety of biotechnology products. USDA was designated the lead agency for regulating genetically engineered crops. FDA was designated to review transgenic crop varieties used for food under the Federal Food, Drug, and Cosmetic Act (FFDCA; US Congress 1958). EPA later clarified its interpretation of the statutes to include "plant pesticides" for regulation under the Federal Insecticide, Fungicide, and Rodenticide Act (FIFRA; US Congress 1947) and FFDCA (EPA 1994a, c). NEPA is applicable to all federal agencies. The current regulation of biotechnology products is shown in table 1.2. Not all the products in table 1.2, namely plant-pesticides, were discussed in the original coordinated framework.

Shortly after the coordinated framework was developed, USDA reviewed and approved transgenic crop varieties for field trials under the Federal Plant Pest Act (FPPA; US Congress 1957).[5] According to the act, a plant pest is:

> any living stage of ... insects, mites, nematodes, slugs, snails, protozoa, or other invertebrate animals, bacteria, fungi, other parasitic plants or reproductive parts thereof, viruses, or any organisms similar to or allied with any of the foregoing, or any infectious substances, which can directly or indirectly injure or cause disease or damage in any plants or parts thereof, or any processed, manufactured, or other products of plants.

Using the coordinated framework as a guide, USDA was the first agency to propose a regulation for the review of plants genetically modified with rDNA methods. On June 16, 1987, a Federal Register notice established procedures for obtaining permits for releasing genetically engineered organisms into the environment in field trials (USDA 1987). Under that regulation, coverage extends to organisms or substances that meet the definition of a plant pest or that USDA chooses to designate as a plant pest.

---

[5]The FPPA supplements and extends the much older Plant Quarantine Act.

TABLE 1.2   Summary of the Current Regulation of Biotechnology
Products, as currently described on the USDA website for Regulatory
Oversight in Biotechnology

Agency	Jurisdiction	Laws
US Department of Agriculture	Plant pests, plants, veterinary biologics	Federal Plant Pest Act
Food and Drug Administration	Food, feed, food additives, veterinary drugs, human drugs, medical devices	Federal Food, Drug, and Cosmetic Act
EPA	Microbial and plant-pesticides, new uses of existing pesticides, novel microorganisms	Federal Insecticide, Fungicide, and Rodenticide Act; Federal Food, Drug, and Cosmetic Act; Toxic Substances Control Act

Source: USDA 1999e.

1.4.3   The First Field Trials

In November and December 1987, USDA issued permits for three
engineered herbicide-tolerant varieties of tomato (two from DuPont and
one from Calgene) and two herbicide-tolerant varieties of tobacco (from
Calgene). These plants were tolerant of the herbicides glyphosate, bromoxynil, or sulfonylurea. Tolerance was based on the overexpression of
the herbicide target in the tolerant plant, the expression of resistant forms
of the target enzymes, or the expression of enzymes that could degrade
the herbicide (Comai et al. 1985; Harrison et al. 1996).

Transgenic pest-protected plants were developed in parallel to herbicide-tolerant plants. The first transgenic pest-protected plant was engineered to contain a coat-protein gene from the tobacco mosaic virus (TMV)
(Powell-Abel et al. 1986); the gene confers resistance to TMV itself, and to
viruses similar to TMV. A transgenic TMV-resistant tomato line developed by Monsanto was approved for field trials on March 23, 1988.

Today, a large portion of US corn and cotton acreage is planted with
transgenic pest-protected plants (Economic Research Service 1999a, b;
Carozzi and Koziel 1997). Those transgenic pest-protected plants contain
genes from the bacterium *Bacillus thuringiensis* (Bt). Bt produces several
proteins during sporulation including endotoxins. Upon ingestion by an
insect, the protoxin form of endotoxin undergoes cleavage in the insect
gut to a truncated active form, which kills insects by binding to receptors

in the insect gut and forming pores (Gill et al. 1992). The pores cause the gut contents to leak into the blood, and this eventually leads to insect death. About 60 proteins from more than 50 subspecies of Bt have been identified in the last 20 years (Federici 1998). Which insects Bt toxins affect depends on the class of Bt protein; they include moths and butterflies (lepidopterans), flies and mosquitoes (dipterans), and beetles (coleopterans).

Mixtures of Bt have been used to spray crops for over 50 years. However, the Bt toxin generally loses its effectiveness in the environment within a few days. Sometimes spraying needs to be done frequently. In transgenic crops, Bt toxin is continuously produced and is protected from the elements. It therefore retains its ability to kill pests during the entire growing season. Moreover, the toxin is generally expressed in every part of the plant, including internal tissues that are difficult to protect with topically applied pesticides. This internal production provides protection against pests that are internal feeders such as the pink bollworm in cotton and the European corn borer in corn. On the other hand, the constant presence of Bt toxin in transgenic pest-protected plants during the growing season has led to concerns about its persistence in the environment and increased probability of pest evolution to overcome the protection mechanism.

The first permit to field-test transgenic crops that contained Bt genes was issued to Monsanto in 1988 for tomato. Initial attempts to make crops that would resist pests in the field were not successful because of problems with the expression of the bacterial genes. In 1990, the first successful Bt crop, cotton (*Gossypium hirsutum*), was produced by overcoming translational and transcriptional barriers to bacterial-gene expression in plants (Perlak et al. 1990). Transgenic methods for introducing Bt are often followed by conventional breeding with varieties that express other useful agronomic traits.

Before transgenic crops were commercialized (from 1987 to 1994), the USDA approved field trials of nine nematode-resistant transgenic pest-protected plant varieties, 45 fungus-resistant varieties, 17 bacteria-resistant varieties, 322 insect-resistant varieties, and 194 virus-resistant varieties.

1.5 AGENCY POLICIES REGARDING COMMERCIALIZATION OF TRANSGENIC PEST-PROTECTED PLANTS

1.5.1 USDA Policy

In 1992, four years after the first field trials began, USDA proposed a regulation that described a petition process for determining that particular plants would no longer be regulated and therefore could be commercially planted. The regulation was finalized in 1993 (USDA 1993). For a crop to achieve nonregulated status, "environmental assessment" and "determination of nonregulated status" documents are prepared by USDA; the documents address safety concerns under the FPPA such as impacts on agriculturally beneficial organisms, as well as addressing the agency's NEPA requirements (section 4.1.1).

1.5.2 FDA Policy

Also in 1992, FDA published a policy statement on its role under FFDCA for reviewing new plant varieties developed by all methods, whether transgenic or conventional (FDA 1992) (section 4.1.2). The FFDCA authorizes FDA to control foods that are "adulterated" with added substances, including naturally occurring substances. The 1992 policy established FDA's role in reviewing the overall composition of the nutrients and toxicants in genetically modified plants. The policy states that "key factors in reviewing safety concerns should be the characteristics of the food product, rather than the fact that the new methods are used."

Under the 1992 policy, FDA asks that companies develop information to determine whether or not the company is obligated to come to the agency for formal regulatory review. Considerations include genetic stability, compositional and nutritional quality attributes of the plant, and toxicity and allergenicity of the gene product. FDA requires that companies submit nutritional and safety data to the agency if there is reason to believe that new plant varieties may pose risks. After publishing its 1992 policy, FDA recommended that companies developing transgenic varieties consult with the agency before marketing a new variety. Guidelines for this voluntary consultation process were published by the FDA in October 1997 (FDA 1997c). Over 45 transgenic plants, including numerous transgenic pest-protected plants and all crop plants that have been marketed in the United States, have gone through the consultation process. FDA has not required that any of the proteins added to transgenic plants be reviewed as food additives[6].

1.5.3 EPA Policy

In the early 1990s, EPA held numerous public meetings of the Biotechnology Science Advisory Committee (BSAC) to develop its regulations for products of biotechnology. In 1994, EPA published proposed

---

[6]However, the antibiotic resistance genes in the Flavr Savr tomato were reviewed as food additives at the request of the tomato's manufacturer (section 1.5.4).

rules that help clarify the agency's role in the coordinated framework by describing which plant-pesticides would be regulated under FIFRA and FFDCA and which would be exempt from regulation (section 4.1.3). Although the proposal has not been finalized, the agency has been implementing its essential elements in registering or exempting plant-pesticides since 1995. EPA defined a plant-pesticide as "a pesticidal substance produced in a living plant and the genetic material necessary for the production of that pesticidal substance, where the substance is intended for use in the living plant." The genetic material necessary for production of a pesticidal substance was included in the definition of plant-pesticide to enable regulatory coverage, under FIFRA, of plant parts such as seeds and pollen where the pesticidal substance might not be expressed. However, with regard to regulation under FFDCA, the agency proposed to establish a categorical exemption from the requirement of a tolerance for this genetic material.

EPA's proposed FIFRA regulation establishes three categories of plant-pesticides that would be exempt from regulation under FIFRA. The first category contains plant-pesticides whose genetic material encodes for a pesticidal substance that is derived from plants that are sexually compatible. The second category of plant-pesticides that are exempt from FIFRA regulation are those that act by affecting the plant so that the target pest is inhibited from attaching to and/or invading the plant tissue by, for example, acting as a structural barrier, or by inactivating toxins produced by the target pest. The third category consists of substances that are coat proteins of plant viruses.

Substances that are exempt from regulation under FIFRA are not automatically exempt from the requirement of a tolerance under FFDCA. Therefore, EPA proposed three additional regulations under FFDCA to accomplish this. Similar to the FIFRA exemption, pesticidal substances derived from plants that are sexually compatible are proposed to be exempt from the requirement for a tolerance. In addition, pesticidal substances derived from plants that are not sexually compatible would also be exempt from the requirement of a tolerance provided that the following two conditions are met: (1) the genetic material encoding the pesticidal substance is derived from a food plant; and (2) the pesticidal substance does not result in a new or significantly different human dietary exposure. EPA also proposed to exempt coat proteins of plant viruses and, as noted earlier, to exempt the genetic material that encodes pesticidal substances.

In 1997, EPA published supplemental notices of proposed rulemaking for the FIFRA and FFDCA proposals published in 1994. EPA took this action in order to allow the public to comment on the Agency's evaluation of the requirements imposed by the Food Quality Protection Act of 1996

(FQPA) (Public Law 104-170, EPA 1997b) that the agency did not address in the 1994 proposals. FQPA amended FFDCA and FIFRA to include a new safety standard for pesticide residues on food, one notable change being special safety factors for children.

1.5.4   Commercialization of Transgenic Pest-Protected Plants

Under the above USDA, FDA, and EPA statutes, the first transgenic crop varieties were approved for commercial planting in the early 1990s. In 1992, the first transgenic crop variety achieved nonregulated status from USDA. This variety was a tomato line for altered fruit ripening developed by Calgene (Flavr Savr). In addition to USDA review, FDA reviewed the safety and nutritional aspects of the Flavr Savr tomato and a food additive petition from Calgene for the use of the kanamycin resistance trait in tomatoes, cotton, and canola (Kahl 1994). Since this review, FDA conducts its assessments for genetically engineered crops by consulting with companies about the safety and composition of the variety and has not required a food additive petition for any other transgenic product, although it could make such a request in the future. EPA was not involved in reviewing the Flavr Savr tomato because the transgenic modification of the tomato did not involve a pesticidal trait.

In December 1994, the first transgenic pest-protected plant achieved nonregulated status from USDA: a virus-resistant squash variety developed by Upjohn/Asgrow Seed Company that contained watermelon mosaic virus-2 coat protein and zucchini yellow mosaic virus coat protein. The USDA assessments for this crop address such concerns as the likelihood of creating new plant viruses via recombination of the introduced coat-protein gene with naturally occurring viruses, the potential of the two new virus-resistance genes to cause squash to become a weed, and the movement of the genes to wild squash relatives. EPA also reviewed this crop. In the July 27, 1994, *Federal Register*, EPA published a notice that Asgrow Seed Co. had submitted a pesticide petition to EPA under FFDCA to exempt the coat proteins from the requirement of a tolerance (EPA 1994b). EPA reviewed the petition for safety concerns, such as toxicity, and established an exemption from the requirement of a tolerance under FFDCA for "residues of the plant-pesticides, as expressed in Asgrow line ZW20 of *Cucurbita pepo* L. and the genetic material necessary for the production of these proteins." EPA also proposed to exempt viral coat protein genes and gene products from review and registration under FIFRA (Section 1.5.3) (EPA 1994a and 1997b).

Varieties employing the Bt resistance mechanism were the next pest-protected plants to achieve nonregulated status from USDA and to have their gene products reviewed as plant-pesticides by EPA. A Bt potato line

TABLE 1.3   Plant Pesticides Reviewed by EPA

Protein	Source
Cry9C[a]	*Bacillus thuringiensis*
Cry1A(b)[b]	*Bacillus thuringiensis*
Cry1A(c)[b]	*Bacillus thuringiensis*
Cry3A[c]	*Bacillus thuringiensis*
viral coat protein	cucumber mosaic virus
viral coat protein	papaya ringspot virus
viral coat protein	virus Y
viral coat protein	watermelon mosaic virus
viral coat protein	zucchini yellow mosaic virus
replicase	potato leaf roll virus

[a]Reviewed for use in corn.

[b]Reviewed for use in all plants.

[c]Reviewed for use in potatoes.

Source: EPA 1999a

resistant to beetles was developed by Monsanto and was cleared for commercial release by USDA in March 1995, subject to EPA and FDA review. The Cry3A delta-endotoxin from Bt was reviewed by EPA in early 1995. An exemption under FFDCA from the requirement of a tolerance for this Bt toxin and the genetic material necessary for its production eliminated the need to establish a maximal permissible level for residues of this Bt toxin in potatoes. For the exemption, EPA reviewed data on toxicity and allergenicity and convened a subpanel of the FIFRA Scientific Advisory Panel to discuss its review; the panel concluded that the Bt potato presented "little potential for human dietary toxicity." Table 1.3 lists the plant pesticides that have been reviewed by EPA.

1.5.5   Current Profile of Transgenic Plants

Over 40 transgenic crop varieties have been cleared through the federal review processes for commercial use in the United States. Of them, 17 (as of December 1999) contain transgenes for pest-protection. Of the 17, 14 containing Bt genes have been developed and cleared by USDA for commercial release (table 1.4) (USDA 1999b). Although the EPA 1994 rule is not yet final, the plant-pesticides in these crops have been reviewed and their gene products registered as plant-pesticides by EPA (table 1.3). Five virus-resistant transgenic pest-protected plant varieties have achieved nonregulated status from USDA (table 1.4).

Transgenic crops were first planted commercially in the 1995 growing season. Since then, their use has been rapidly increasing. In 1997, 20.3

TABLE 1.4 Crops Deregulated by the USDA with Transgenic Pesticidal Traits

Crop	Resistance[a]	Company	Date of Nonregulated Status
Insect resistance			
Potato	Bt IR	Monsanto	March 1995
Corn	Bt IR	Ciba-Geigy	May 1995
Cotton	Bt IR	Monsanto	June 1995
Corn	Bt IR	Monsanto	August 1995
Corn	Bt IR	Northrup King	January 1996
Corn	Bt IR	Monsanto	March 1996
Potato	Bt IR	Monsanto	May 1996
Corn	Bt IR	Dekalb Genetics	March 1997
Cotton	Bt IR, HR	Calgene	April 1997
Corn	Bt IR, HR	Monsanto	May 1997
Tomato	Bt IR	Monsanto	March 1998
Corn	Bt IR, HR	AgrEvo	May 1998
Virus and insect resistance			
Potato	Bt IR, VR	Monsanto	December 1998
Potato	Bt IR, VR	Monsanto	February 1999
Virus resistance only			
Squash	VR	Upjohn/Asgrow	December 1994
Squash	VR	Asgrow	June 1996
Papaya	VR	Cornell University	September 1996

[a]Bt IR = Bt endotoxin-based insect resistance; VR = virus resistance; HR = herbicide resistance

Source: USDA 1999b.

million acres of transgenic crops were planted in the United States; in 1998, 50.2 million acres were planted (James 1998); and in 1999, 70 million were planted (James 1999). A total of 98 million acres were planted worldwide in 1999 (James 1999). Transgenic pest-protected crop varieties that contain Bt toxin transgenes make up a large percentage of the commercial transgenic crops. In 1998 in the United States, about 25% of total cotton acreage and 21% of total corn acreage were planted with transgenic crops that contain Bt genes (USDA 1999d). In 1999, approximately 30 million acres of insect protected crops were planted in the US (James 1999). So far, many of these transgenic pest-protected crops seem to be effective in controlling pests; a reduced need for chemical pesticides and increased yields have been reported by many, but not all, growers (Robinson 1998; Gianessi and Carpenter 1999). For example, one report indicates that insecticide sprays in cotton were reduced in 1998 from an average of 8.3 insecticide applications for conventional cotton to an average of 6.0 sprays for Bt cotton (Mullin and Mills 1999), which led to an estimated reduction of over 5 million acre-treatments and over 2 million pounds of chemical insecticide (Gianessi and Carpenter 1999). However, estimated benefits might depend on the baseline level of pest infestation during a specific growing season and on the techniques used to make comparisons (USDA 1999d). The use of transgenic pest-protected crops has been profitable in growing regions subject to severe pressure from specific pests or where alternative means of pest control have been infeasible or expensive. For example, Bt cotton has been accepted by a large percentage of growers in states where pest resistance to synthetic pyrethroids has left them without chemical means of controlling bollworms, but limited in other regions where pest hazards are not so extreme (USDA 1999d; Falck-Zepeda et al. 1999). Adoption of Bt corn has similarly been limited to areas with the highest pest pressure (USDA 1999d).

In addition to the approved commercial transgenic crop varieties, thousands of transgenic varieties are undergoing field trials (USDA 1999c). From 1987 through January 2000, the number of permits issued and notifications acknowledged was over 6700; about 3000 were for varieties having pest-resistance genes (table 1.5).

1.6 PUBLIC CONCERNS AND ISSUES

Given the rapid increase in acres planted with commercial transgenic crops and the likely additional increase in their use, many groups have raised concerns about the ecological and human health risks that might be posed by these crops (Ho 1998). Although the risks might not, in principle, differ in type from those associated with other products (for example, conventional pest-protected plants, pesticides), the public has focused its attention on transgenic crops.

Concerns over pesticidal traits include the enhanced evolution of resistant pest strains, the toxicity or allergenicity of the gene products to

TABLE 1.5   Number of Permits Issued for or Notifications of Field Trials in the United States Involving Crops with Pest-Resistance Genes, 1987-1999.

Category of Resistance	Number (% of all field trials)
Insect	1505 (22)
Virus	1013 (15)
Fungal	378 (5.5)
Bacterial	78 (1.2)
Nematode	7 (0.1)

humans, the hybridization of transgenic pest-protected plants with neighboring wild relatives, and adverse effects on nontarget organisms. These concerns are presented below and discussed more extensively in chapters 2 and 3 where the scientific bases and empirical evidence are analyzed.

1.6.1   The Development of Pest Resistance to Engineered Traits

Farmers and gardeners who use microbial Bt sprays are concerned that the widespread commercial planting of transgenic pest-protected plants with Bt genes will lead to rapid development of insect resistance to Bt, which will in turn make their microbial sprays ineffective. Instances of pest adaptation to conventional Bt products have been documented (Tabashnik et al. 1994).

Scientists who conduct research on pest resistance to plant-protection mechanisms published resistance management strategies for Bt corn, cotton, and potato (McGaughey and Whalon 1992; Tabashnki 1994; Roush 1997; Gould 1998; UCS 1998), and the EPA published findings of a specially convened scientific advisory panel on Bt resistance management (SAP 1998). Under the registration process for plant pesticides, EPA requires a particular amount of non-Bt cotton or corn to be planted next to Bt cotton or corn to serve as a refuge for insects carrying Bt susceptible genes, and they also encourage the development of resistance management strategies for other transgenic Bt crops. However, the percentage of acreage that is needed to provide a sufficient refuge to avoid the rapid development of pest resistance and the proper location of the refuge are debated by industry, entomologists, and environmental groups (Inside EPA 1999; UCS 1998) (see section 2.9). Recently, the EPA placed new restrictions on growing transgenic Bt corn which include a requirement that farmers plant 20% to 50% of their corn acreage with conventionally bred corn (EPA 1999h; Weiss 2000).

1.6.2   Human Health Concerns

Allergenicity due to transgenic gene products has been highlighted as a human health concern (Metcalfe et al. 1996a, b) (see section 2.5.1). Guidance for assessing these concerns was provided in a 1996 report published by the International Food Biotechnology Council in conjunction with the International Life Sciences Institute (Metcalfe et al. 1996b). One transgenic plant was shown to have allergenic properties during laboratory tests (Nordlee et al. 1996). To improve the nutritional quality of soybeans, a transgenic plant containing a methionine-rich protein from Brazil nuts (*Bertholletia excelsa*) was developed by Pioneer Hybrid International. The company discontinued development of this product as a result of these allergenicity concerns. It is important to note that modern biotechnology can also be used to reduce the allergenic risks associated with our current food supply. For example, Matsuda et al. (1998) have published papers showing that they reduced the major allergen in rice by approximately 80% by using antisense rDNA technology.

1.6.3   Gene Flow and Cross Pollination With Weedy Relatives

Other safety issues which have received attention are those involving ecological risks such as the effects of gene flow. Studies have been conducted to assess the potential for gene flow among and within related species (see sections 2.7 and 3.4.1). The ability of transgenic plants to cross-pollinate with their wild relatives and form offspring with enhanced weediness has been investigated when herbicide-tolerant rapeseed plants were back-crossed with a wild relative. The hybrid progeny plants produced an equivalent amount of seed as the wild genotypes and were also herbicide-resistant (Snow et al. 1999). That study indicated that back-crossed generations of hybrids between transgenic and nontransgenic crops can have the same potential to flourish as other plants. In a more controversial study, wild type *Arabidopsis thaliana* plants were found to be fertilized by pollen from transgenic plants more often than by pollen from nontransgenic plants (Bergelson et al. 1998). In addition to those experiments, the use of models has been explored to assess the invasiveness of engineered organisms, although indications are that these models will require several years worth of data to be validated (Kareiva et al. 1996).

1.6.4   Nontarget Species

Although some transgenic pest-protected plants have the potential to reduce pesticide use and thus to prevent substantial environmental damage, there is concern that gene products from the plants could harm beneficial insects or birds (nontarget species) that are in direct contact with the plants or that feed on insects that are (see section 2.6 and 3.1.2). Hillbeck et al. (1998a, b) found that when chrysopid larvae were reared on prey that were fed Bt-producing corn, they had 62% mortality. When they were reared on prey that were fed non-Bt corn, mortality was only 37%.

Another experiment indicated that Bt toxins can bind to humic acids from soil, be protected from biodegradation, and persist in the soil (Crecchio and Stotzky 1998). It is not known whether nontarget organisms would be affected by bound toxin molecules in field situations. Other studies indicate that Bt toxins generally degrade quickly in the soil (Palm et al. 1994; Sims and Sanders 1995; Palm et al 1996).

A well-publicized recent laboratory study indicated that when monarch butterfly larvae were fed milkweed dusted with transgenic Bt pollen, high mortality was exhibited (Losey et al. 1999). The relationship between this preliminary laboratory finding and field effects is unclear (Yoon 1999). One recent field test reports that at least 500 pollen grains per square centimeter is necessary to sicken monarch caterpillars and that milkweed plants growing adjacent to corn fields had only an average of 78 grains per square centimeter (Kendall 1999) (see section 2.6.2). In other experiments, however, monarch caterpillars that consumed concentrations of Bt corn pollen (Event 176) naturally deposited on milkweeds in the field experienced 20% mortality with only 48 hours of exposure (Hansen and Obrycki 1999a,b). Further field-based research is needed to determine whether dispersed Bt pollen could have detectable effects on the population dynamics of nontarget organisms.

1.6.5   Regulatory Concerns

The above concerns have led some to question the safety review that transgenic crops receive in the United States under the coordinated framework. Many believe that transgenic crops present substantial human health and ecological risks, and that these are not properly assessed by the regulatory framework. But many others believe that the risks are minimal, that the benefits outweigh the risks, and that the current regulatory scheme is perhaps onerous.

Cited benefits include a reported 250,000-gallon reduction in chemical pesticide use in 1996 and a 30-50% reduction in the number of insecticide applications over the period of 1996-1998 due to the growing of commercial transgenic Bt cotton (Robinson 1998; Williams 1997, 1998, and 1999). The reduction might prevent much environmental damage. In addition, Bt toxins have specific insect targets, whereas traditional broad-spectrum chemical insecticides often kill insects more indiscriminately (Federici 1998). This may lead to outbreaks of secondary pests requiring the use of more insecticides. However, many believe that transgenic pest-protected plants should not only be compared to the use of chemicals, but also to alternative methods such as biological control.

The debate has intensified in recent months, given the international concerns and impending regulatory decisions in the United States. In March 1999, Congress held a hearing on the 1994 proposed EPA plant-pesticide rule (Hart 1999c). Although transgenic pest-protected plants have been registered under this rule in the last 5 years, the rule has not been finalized, and its scientific and legal validity are being questioned. The EPA planned to finalize the rule by the end of 1999. The debate over this rule has many facets. Environmental and consumer groups argue that the EPA is not rigorous enough in its scientific review (Hart 1999c) and that the proposed rule has too many exemptions. They are also concerned that the EPA rule does not adequately cover all of the risk issues. Several professional societies have argued that EPA is overstepping its boundaries by reviewing plant gene products as pesticides, stating that this could damage the progress of science by overburdening small biotechnology companies and public breeding programs with the cost of regulation, as well as undermining confidence in the food supply (Eleven Scientific Societies 1996; CAST 1998). Some congressional members are concerned about the lack of a formal cost-benefit analysis to accompany the rule and about whether the definition of a pesticide in FIFRA gives EPA the authority to regulate transgenic pest-protected plants (Hart 1999c).

Given the debate about its proposed rule, EPA held a workshop in 1997 to address some of the criticisms (EPA 1997c) and is incorporating changes into the rule on the basis of comments. One comment that is being considered suggests changing the terminology to avoid the use of "plant-pesticides" for gene products of transgenic pest-protected plants. EPA has sought input on a more appropriate name for these traits in a recent *Federal Register* notice (EPA 1999c). A change might address the public's concern about labeling plants as "pesticides"; however, it would not address other concerns, such as EPA's authority, its role in the coordinated framework, and whether the risks are being properly addressed by this framework.

1.7 ROLE OF THIS REPORT

In the past, the National Academy of Sciences (NAS) and National Research Council (NRC) have had the opportunity to provide guidance to scientists, regulatory agencies, and the public concerning rDNA issues. The 1974-1975 efforts helped to initiate the national debate over the safety of genetically engineered organisms (Berg et al. 1974). In 1987, given the proposed release of genetically engineered organisms into the environment, the NAS Council issued a white paper, *Introduction of Recombinant DNA-Engineered Organisms into the Environment* (NAS 1987), which proposed guiding principles that helped shape national policy for the review of genetically engineered organisms. In 1989, the NRC convened a committee to establish a framework for decisions regarding the field testing of genetically engineered organisms (NRC 1989); the criteria and methods for evaluation suggested by that committee have been guiding USDA oversight of field trials for transgenic crops in the last 10 years. Given the current political and social climate, the NRC believes that it has a role to play in addressing the scientific issues surrounding the regulation of transgenic pest-protected plants.

The scope of the study and structure of this report are outlined in the executive summary (section ES.1). Transgenic pest-protected plants[7] are the focus of the committee's discussion of the regulatory framework, inasmuch as the framework is designed for transgenic plants (OSTP 1986). Conventional pest-protected plants are discussed for scientific comparisons. Given impending decisions with respect to the EPA plant-pesticide rule, the committee focused on the EPA's proposed rule, but also addressed the roles of the EPA, USDA, and FDA under the coordinated framework.

The committee hopes that this report will provide guidance for reviewing the thousand or more transgenic pest-protected plants that are being tested in the field as well as those yet to be developed. Although transgenic Bt crops have received the most attention given their commercial use, the committee proposes to look towards the future by discussing general issues concerning transgenic pest-protected plants for which there may be fewer data and that could have an impact in coming years. It is not possible for the current committee to comment on other classes of transgenic crops (such as herbicide-tolerant crops) given the breadth of the issues and the time frame; however, some of the conclusions in this report regarding transgenic pest-protected plants might be applicable to other transgenic crops and are indicated as such. Terms that frequently appear in the report are defined in the executive summary (ES.3), a list of acronyms can be found in appendix D, and common and scientific names for the various organisms listed in the text appear in appendix E.

---

[7]Note that the committee focused on potential impacts of food and fiber crops, not on the potential impacts of other types of transgenic pest-protected plants that might be commercialized in the future (for example, forest trees).

2

Potential Environmental and Human Health Implications of Pest-Protected Plants

This chapter begins with a discussion of risk assessment and its application to pest-protected plants and includes a review of the 1987 National Academy of Sciences white paper. It then considers the array of pest-protection traits and their possible use in transgenic pest-protected plants. The bulk of the chapter discusses potential environmental and human health impacts of conventional and transgenic pest-protected plants, such as human toxicity and allergenicity, nontarget effects, hybridization with weedy relatives, and evolution of pest adaptation to pest-protected plants. Scientific data on the potential for adverse environmental and health effects are presented and discussed. Scientific review in federal agencies is also discussed and will be covered in more detail in chapter 3.

2.1 RISK ASSESSMENT AND PEST-PROTECTED PLANTS

The 1987 National Academy of Sciences (NAS) white paper *Introduction of rDNA-Engineered Organisms into the Environment* stated that the "risks" posed by transgenic organisms are the "same in kind" as those associated with the introduction of unmodified organisms and organisms modified by other methods. Similar conclusions have been reached by international scientific organizations (FAO/WHO 1996; OECD 1993 and 1997). A clear definition of risk is needed if the committee is to interpret and evaluate that statement appropriately. This section clarifies the meaning of risk and related terms according to well-accepted definitions (NRC 1983).

Risk assessment consists of four steps: hazard identification, dose-response evaluation, exposure assessment, and risk characterization.[1] The definitions of those and other terms in the National Research Council's (NRC's) "Red Book" (NRC 1983) are widely used and generally accepted.

Hazard identification is "the determination of whether a particular chemical is or is not causally linked to particular health effects" (NRC 1983). Hazard is usually determined experimentally in controlled experiments with known doses. In the case of pest-protected plants, hazard would be the effect of a gene product (such as *Bacillus thuringiensis* (Bt) toxin, or a secondary plant product, such as a glycoalkaloid) which is expressed or changed as a result of genetic modification. The effects of gene flow or the effects on nontarget organisms could be considered potential hazards for ecological risk assessments.

Dose-response assessment is the determination of the relationship between the magnitude of exposure and the probability of occurrence of the adverse effect in question. Dose-response assessment can address the potency or severity of the hazard. For example, many substances lead to adverse effects only at high doses and might be regarded as posing less severe hazards. The relationship between dose and adverse effects for a particular hazard is reflected in the dose-response curve. In the case of pest-protected plants, some proteinase inhibitors require very high concentrations to cause adverse health effects (Ryan 1990). On the other hand, some plant glycoalkaloids cause adverse health effects at relatively low doses. This allows toxicants to be ranked according to "relative hazard" which is not the same as "relative risk." Overall risk is the product of the likelihood of an adverse consequence and the severity of that consequence. Hazard severity, and probability and magnitude of exposure all contribute to the overall risk. The risks that may be posed by proteinase inhibitors and glycoalkaloids could be similar depending on the probability and magnitude of exposure.

Exposure assessment is the determination of the extent of exposure to a toxicant under any stated set of circumstances. In the context of pest-protected crops, exposure of nontarget species to a plant-pesticide might be considered for ecological risk assessment, and exposure of humans to a plant-pesticide for human health risk assessment. Exposure assessment of pest-protected plants should deal with such questions as how much of the toxicant humans consume, concentrations in the edible portions of the crop, and how often and how much nontarget insects consume.

Risk characterization considers all the above and is often reported as a quantitative assessment of the probability of adverse effects under de-

---

[1]Note that these essential steps may be categorized and/or termed differently in various risk assessment frameworks.

fined conditions of exposure—for example, one in 10,000 humans will become ill given a certain set of circumstances. Hazard identification, dose-response assessment and exposure assessment are all essential elements of a risk assessment.

Standard toxicological human health risk assessment, despite problems of uncertainty and variability and the consequent difficulty in extrapolation, is science-based. Variability is the range of differences implicit in a natural population (such as the genetic variability in sensitivity to allergens); uncertainty is based on incomplete knowledge or data (such as inadequate surveys of genetic variability to allergens) or on measurement error.

Quantitative risk assessment is being used for not only cancer or toxicological risk assessment, but also for ecological risk assessment, microbial risk assessment, and other diverse types of assessment. In principle, quantitative risk assessment of transgenic pest-protected plants could be based on the methods of quantitative risk assessment if a hazard is detected. If adequate data were not available, the assessment could use uncertainty analyses, ranges of values, and extrapolation. However, until methods are adapted and applied to quantitative risk assessments for pest-protected plants, "relative hazard" ranking may be the best approach, recognizing that this is an interim solution and that quantitative risk assessment is the desired goal.

Because the fundamental elements of risk assessment, such as hazard identification, dose-response assessment, exposure assessment, and risk characterization, can also be applied to risk assessments for transgenic pest-protected plants, the committee found that

Health and ecological risk assessments of transgenic pest-protected plants do not differ in principle from the assessment of other health and ecological risks.

2.2   REVIEW OF PREVIOUS NATIONAL ACADEMY OF SCIENCES AND NATIONAL RESEARCH COUNCIL REPORTS

2.2.1   Introduction of Recombinant DNA-engineered Organisms Into the Environment (1987)

In 1987, the NAS published a summary of key issues related to the introduction of recombinant DNA-engineered (rDNA-engineered) organisms into the environment (NAS 1987). This brief white paper outlined the expected risks and benefits associated with all types of transgenic organisms, including bacteria, insects, fish, and crop plants. At the time, commercial field releases of transgenic organisms were still in the planning stages, and the impending "biotechnology revolution" attracted enthusiastic support from some quarters and strong disapproval from others. To address the perception that rDNA techniques might be intrinsically dangerous, the report offered the following conclusions:

- point 1 "There is no evidence that unique hazards exist either in the use of rDNA techniques or in the movement of genes between unrelated organisms."
- point 2 "The risks associated with the introduction of rDNA-engineered organisms are the same in kind as those associated with the introduction of unmodified organisms and organisms modified by other methods."
- point 3 "Assessment of the risks of introducing rDNA-engineered organisms into the environment should be based on the nature of the organism and the environment into which it is introduced, not on the method by which it was produced."

Throughout this report, the committee describes various methods of both conventional and transgenic breeding methods in detail to provide relevant information about their similarities and differences. Some of the similarities and differences in properties of plants produced by varied genetic approaches are presented in box 2.1. Properties of conventional pest-protected plants are discussed, but the committee focuses on risks and benefits that may be posed by growing transgenic pest-protected plants commercially and on their regulatory oversight under the coordinated framework for regulation of genetically engineered organisms.

The 1987 NAS report noted that the risks associated with rDNA-engineered organisms are "the same in kind" as those associated with unmodified organisms and organisms modified by other methods. The committee agrees with that statement for pest-protected plants in that both transgenic and conventional plants may pose certain risks and the resulting plant phenotypes are often similar. Transgenic breeding techniques can be used to obtain the same resistance phenotype as conventional methods (for example resistance to microbial pathogens, nematodes, and insects). Because both methods have the potential to produce organisms of high or low risk, the committee agrees that

The *properties* of a genetically modified organism should be the focus of risk assessments, not the *process* by which it was produced (point 3).

In this regard, the committee found that

There is no strict dichotomy between, or new categories of, the health and environmental risks that might be posed by transgenic and conventional pest-protected plants.

BOX 2.1
Summary of Genetic Basis of Resistance Traits That Have Been Bred into Cultivated Plants Using Conventional and Transgenic Techniques

1) Conventionally bred plants only
a) Polygenic traits[2] (controlled by several interacting genes, usually selected without knowledge of which genes are involved)

2) Both Conventionally bred and transgenic plants
a) Single-gene traits[2] from the same species or a related species
b) Several single-gene traits that are not genetically linked and are therefore inherited independently
c) Several single-gene traits that are physically linked and inherited as a unit; occasionally possible with conventional breeding, as when a chromosome segment bearing more than one resistance gene is transferred to the cultivar usually accompanied by extraneous DNA; transgenic methods allow several single-gene traits to be tightly linked without extraneous DNA
d) Single-gene traits expressed only in particular tissues or at particular developmental stages because of specific promoters; occasionally possible with conventional breeding, but more flexible and precise with transgenic methods

3) Transgenic plants only
a) Single-gene traits found in the same species or a related species and modified by changes in the nucleotide sequence of the structural gene or the promoter to improve the plant's phenotypic characteristics
b) Single-gene traits obtained from unrelated organisms (such as viruses, bacteria, insects, vertebrates, and other plants); sometimes modified by a change in the nucleotide sequence of the structural gene or the promoter to improve the plant's phenotypic characteristics
c) Single-gene traits that can be induced by a chemical spray or by specific environmental conditions (such as threshold temperature), based on the action of specific promoters; (these traits may also occur naturally in nontransgenic plants, such as those with systemic acquired resistance, but have rarely been selected intentionally by conventional breeding)

The committee recognizes that the magnitude of the risk varies on a product by product basis. The committee also agrees with points 1 and 2 in the sense that the potential hazards and risks associated with the organisms produced by conventional and transgenic methods fall into the

---

[2]A molecular technique known as marker-assisted selection can speed the identification of polygenic or single-gene traits in the plant's own genome, and rapid advances in genomics are expected to speed the identification of additional single-gene resistance traits in plants and other organisms.

same general categories. As this report discusses, toxicity, allergencity, effects of gene flow, development of resistant pests, and effects on non-target species are concerns for both conventional and transgenic pest-protected plants.

The committee agrees with the 1987 NAS principles in that the magnitude of quantitative risk does not depend on the genetic-modification *process*. It depends on the new genes that are expressed in the plant. End points of risk (such as illness in humans and declines in nontarget species) can be the same regardless of whether a specific new gene was transferred by conventional or transgenic methods. For example, if the same alkaloid gene is transferred by sexual hybridization or *Agrobacterium*-mediated insertion, the risk should be similar. If a gene coding for a novel trait is transferred by transgenic methods, but cannot be transferred by conventional methods, it is the expressed trait that requires scrutiny, not the method of transfer. In summary,

The present committee found the three general principles to be valid within the scope of issues considered by the 1987 paper, and the present report further clarifies and expands on these principles.

Throughout the report, the committee expands on the 1987 principles by describing various methods of both conventional and transgenic plant breeding, and their potential consequences. The greater diversity of genes that can be transferred by transgenic methods, their enhanced effectiveness, and the ability to insert the same gene into many cultivated species have led to concerns about transgenic crops. Does the potential of transgenic methods to expand on the diversity of transferred genes mean that there is a greater chance for unintended risks from transgenic plants than those from conventionally bred plants? That question has been the subject of considerable debate and draws the question away from specific products. Some transgenic breeding results in pest-protective traits that are phenotypically indistinguishable from those conferred by conventional methods. In addition, transgenic methods are based on more complete knowledge of the genes that are being transferred into cultivated plants. In other cases, however, transgenic pest-protection traits may result in plants having new phenotypes, such as novel plant-produced toxins that could potentially affect human or animal health, nontarget organisms, or the weediness of crop relatives. Transgenic methods can also introduce extraneous traits when they involve marker genes, such as antibiotic resistance genes.

An up-to-date assessment of potential problems and advantages of transgenic methods is warranted (see section ES.2). Transgenic methods can improve the precision of plant breeding and lead to many advantages over current pest control methods. With careful planning and appropriate regulatory oversight, commercial cultivation of transgenic pest-protected plants is not generally expected to pose higher risks and may pose less risk than other commonly used chemical and biological pest-management techniques. The committee concludes that

A major goal for further research and development of transgenic and conventional pest-protected plants should be to enhance agricultural productivity in ways that also foster more sustainable agricultural practices and enhance the preservation of biodiversity, and decrease the potential for health problems that could be associated with some types of pest-protected plants.

2.2.2   Field Testing Genetically Modified Organisms (1989)

To expand on the general principles outlined above, NRC published a more detailed report on how genetically modified plants and microorganisms should be regulated for small-scale, experimental field tests (NRC 1989). The recommendations proved useful and remain well-founded with regard to how federal agencies regulate field testing of genetically engineered organisms. One important and widely accepted conclusion of the 1989 report is that genetically engineered organisms should be evaluated case by case. The report also describes many of the same issues that apply to large-scale introductions, such as the potential to create weeds or insects that are resistant to Bt insecticides. However, because the 1989 report did not directly address health or environmental risks associated with commercialization, it has limited utility for providing guidelines for regulation of transgenic pest-protected plants.

2.3   FORMS AND MECHANISMS OF GENETICALLY CONTROLLED PEST-PROTECTION

Use of genetically controlled pest-protected germplasm for pest management is widely perceived as providing a number of benefits. First, crop losses or damage can be eliminated or minimized resulting in improvement of both yield and quality. Second, resistant germplasm constitutes a low-input option for pest management that often reduces the need for chemical pesticides and their associated financial costs. Third, by reducing the use of traditional pesticides, pest-protected plants can increase the safety of the food supply and reduce environmental impacts. An example of reduced pesticide use and costs as a direct result of planting conventional pest-protected crops is the case of winter wheat bred for resistance to eyespot disease caused by the fungus *Pseudocercosporella herpotrichoides*. Resistant cultivars, which were introduced in 1988 and are now grown on nearly 1,000,000 hectares in the Pacific Northwest United States, have reduced midseason fungicide treatments to roughly half of that needed with susceptible cultivars (Jones et al. 1995). Estimates from 1994 indicate that genetic protection from eyespot disease reduced growers' production costs by $40 per hectare.

Plants with pest-protection properties can inhibit growth, reproduction, or survival of a particular pest or group of pests, or they may tolerate a pest infestation with minimal or acceptable levels of damage. Pest-protected plants that reduce pest populations can exhibit pest-protection characteristics through structural mechanisms. Trichomes on leaf surfaces, for example, present a structural barrier that reduces feeding activity of some insects. Pest-defense systems can also involve intracellular or biochemical mechanisms. These defense mechanisms can work through the action of preformed defensive compounds, and through induced defensive compounds, reactions, and signaling pathways that are triggered specifically or nonspecifically by an invading pest.

To understand the rationale of current and future directions of transgenic breeding for pest-protection and to assess risks of transgenic pest-protected plants relative to those that may be posed by conventional pest-protected plants, this section reviews mechanisms of conventional and transgenic resistance to insects and pathogens.

2.3.1 Natural Pest-protection Mechanisms

Preformed Chemical Defenses

Plants constitutively produce a variety of antimicrobial or insecticidal chemicals that are known or suspected to provide pest-protection (Mansfield 1983; Rosenthal and Berenbaum 1991). The chemicals are often sequestered in specialized cells or expressed in particular organs. Chemicals having antibiotic or suppressive activities against pathogens and insects include saponins, glycoalkaloids, terpenoids, and phenolic compounds. They can have acute or chronic toxic effects and some compounds can have behavioral effects on insects that reduce insect feeding, reproduction, or colonization. The saponin avenacin A-1, for example, is a glycosylated triterpene that is toxic to fungi by perturbing membrane structure and function (Osbourn 1996). It is found in the roots of some cereals. Avenacin A-1 in oats confers resistance to a number of root-infecting fungal pathogens, such as *Gaeumannomyces graminis*. Like other chemical defenses, avenacin A-1 is effective as an antibiotic in proportion to its accumulation in roots, the inherent sensitivity of the fungus, and the ability of the pathogen to detoxify the compound. Some compounds have relatively broad specificities. Cyclic hydroxamic acids, such as 2,4-dihydroxy-7-methoxy-1,4-benzoxazin-3-one (DIMBOA), have been shown to confer protection against both fungal pathogens and insect pests (Frey et al. 1997).

Although many preformed chemicals, such as avenacin A-1 and DIMBOA, have been shown to provide pest-protection, the great majority of natural plant chemicals that have antibiotic properties in vitro have not been proved to be active defensive compounds in vivo. The array of compounds with potential defensive capability is vast, and it includes a large number of potential animal and human toxins. For example, 49 natural products or metabolites found in cabbage are known toxins in microbial or animal models (Ames et al. 1990a). Additionally, a number of natural products in the food supply do have acute human toxicity; the cholinesterase inhibitors solanine and chaconine in potato are well-documented examples. Ames et al. (1990b) estimated that the typical American consumes such compounds at roughly 1.5 g/day, primarily in fruits and vegetables, but diets rich in fruits and vegetables are associated with lower, not higher, risks of illnesses such as certain forms of cancer and heart disease (NRC 1982). Therefore, there is not necessarily a correlation between consumption of fruits and vegetables containing compounds with toxicity in experimental systems and adverse health effects.

Resistance Genes

Although the term resistance gene is sometimes used to describe any gene that encodes a plant-protection mechanism, it is most commonly applied to a gene that triggers a defense response to a specific pest or pathogen. In this report, these pathogen-specific resistance genes will be referred to as race-specific R genes, or simply, R genes. The more general term, defensive genes, will be used to describe natural plant genes specifying antibiotic or insecticidal factors that have broad specificity. The identification and deployment of R genes have been among the most important factors in the development of high-yielding conventional crop varieties. Genes have allowed the continued cultivation of many crops in areas where virulent pathogens and detrimental pests are common (for example, leaf stem, and stripe rust in wheat) (Knott 1989; Line 1995; McIntosh and Brown 1997). In many cases, the use of R genes has permitted a reduction in reliance on externally applied chemical pesticides (Jones et al. 1995).

Genetic interactions between flax and the flax rust pathogen indicated that many R genes are effective against only particular races of a pathogen (or types of a pathogen with specific virulence properties) (Flor 1971). The races that are suppressed by a given R gene are known to contain specific "avirulence" genes; races that are not suppressed lack a functional corresponding avirulence gene. In at least some cases, pathogen avirulence genes encode proteins that are required for infection of susceptible plant hosts (Kearney and Staskawicz 1990). The "gene-for-gene" concept was proposed to explain the interaction between a plant R gene and a pathogen avirulence gene, and this concept is used in agriculture to develop pest-protected crop varieties that are resistant to damage by pathogen races that have known virulence properties. A feature of race-specific R genes, and one of the major limitations associated with their use, is the occurrence of pathogen races that are unaffected by a given plant R gene; these can be pre-existing races that lack the corresponding avirulence genes or new races that have lost avirulence gene function. Study of numerous R genes isolated over the last few years has shown that many have a common evolutionary origin (Baker et al. 1997). Furthermore, race-specific R genes appear to function by triggering a cascade of molecular signaling and biochemical reactions that arrest pathogen spread at the initial site of infection, regardless of whether a particular R gene specifies resistance to a virus, fungus, or bacterium.

Several other types of disease-resistance genes that do not fit the gene-for-gene concept have also been identified. The *HM1* gene of maize encodes a reductase that inactivates HC toxin, a cyclic tetrapeptide required for virulence of the fungus *Cochliobolus carbonum* race 1 (Johal and Briggs 1992). The recessive *mlo* gene in barley confers resistance to all races of the powdery mildew fungus, *Erisyphe graminis* f. sp. *hordei*, by priming the onset of several defense pathways (Buschges et al. 1997). Polygenic traits that confer quantitative pest-protection can also provide durable protection. Although the basis for this type of pest-protection is not entirely clear, cumulative effects of plant R genes that have been overcome by virulent pathogens might play a role in some systems (Li et al. 1999).

Genes for controlling insect and other invertebrate pests have also been identified and deployed, although they might be less common than plant R genes for viral, fungal, and bacterial pathogens. Some encode enzymes that catalyze synthesis of insecticidal or insect-deterrent compounds, whereas others trigger localized defense responses. Several nematode R genes are chemically related, or sequence-related, to race-specific pathogen R genes (Cai et al. 1997; Milligan et al. 1998); this suggests that the signaling mechanisms leading to resistance to nematode are similar to those for resistance to pathogens. The tomato *Mi* gene for resistance to the root-knot nematode, *Meloidogyne incognita*, also confers resistance to the potato aphid, *Macrosiphum euphorbiae* (Rossi et al. 1998; Vos et al. 1998); thus, some insect resistance genes could have broad specificity.

Induced Resistance Responses

A number of resistance responses by plants are induced by pathogen invasion or insect attack (Hutcheson 1998). The hypersensitive response (HR) results after R-gene-mediated, race-specific recognition of a pathogen. The HR in a natural infection is often limited to relatively few cells around the initial infection site. It can also be triggered nonspecifically by various elicitor compounds, such as fungal cell-wall components. The HR involves a cascade of reactions that result in production of reactive oxygen intermediates, antimicrobial compounds (termed phytoalexins), and degradative enzymes; alteration of cell membranes and cell walls; and ultimately cell death. The result of the HR in infected tissues is usually localized necrosis, inhibition of pathogen growth, and limitation of the disease. The HR can occur in plants that contain race-specific R genes effective against all types of viruses, fungi, and bacteria.

The HR leads to a number of other localized and systemic processes that result in increased generalized resistance to a wide array of pathogens. The systemic-acquired-resistance response results in activation of genes that encode defensive proteins, such as glucanases and chitinases, and antimicrobial biosynthetic pathways throughout the plant (Ryals et al. 1996). Defensive proteins can also be induced during the natural course of development of some plants; for example, pathogenesis-related proteins (such as several chitinases and osmotin) with antifungal activity are the predominant proteins that accumulate in the ripening fruit of grape plants (Salzman et al. 1998).

Insect herbivore activity can lead to a systemic defense response (Ryan 1990). This response can be triggered by biotic damage, such as that caused by chewing insects, or by mechanical damage. Insect feeding on a single leaf can result in production of defensive chemicals in all of a plant's leaves (Rosenthal and Berenbaum 1991). An important component of this wound-induced response is activation of genes that encode proteins, such as proteinase inhibitors, that have insecticidal activity. Proteinase inhibitors prevent digestion of plant material in the insect gut, and so result in starvation. Thus, plants exposed to chewing insects gain resistance to additional insect feeding through the wound response.

Viruses activate a defensive response that resembles post-transcriptional gene silencing (PTGS) (Carrington and Whitham 1998). PTGS response is adaptive in providing a customized antiviral response to each new virus that the plant encounters. Silencing in response to viruses with a RNA-based genetic code involves degradation of the genome itself. For viruses with a DNA-based genetic code, the PTGS results in degradation of the transcription products (mRNA). In either case this results in lower virus accumulation or in recovery of the plant. PTGS response can be manipulated in transgenic plants to confer extreme immunity against viruses (Baulcombe 1996).

2.3.2 Transgenic Pest-protection Mechanisms

Plant biologists and breeders have developed a number of plants that have pest-protection conferred by transgenes. Transgenic pest-protection strategies generally depend on expression of novel genetic resources or transfer of natural plant resistance or defense genes. Transgenic pest-protection based on novel genetic traits involves the introduction of genes that interfere with a specific pest but that are derived from organisms in which the gene's natural function is not that of plant protection. The application of transgenic resistance should be most useful where natural conventional breeding has failed due to lack of resistance genes in sexually compatible plants or due to undesirable agronomic traits in conventional pest-protected crops. For example, the oat *Pc-2* resistance gene which controls crown rust disease caused by the fungus *Puccinia coronata* is coinherited with a trait that confers sensitivity to an unrelated fungal pathogen, *Cochliobolus victoriae*, so it would not be useful to deploy this gene in oat cultivars by conventional breeding methods (Walton 1996). Transgenic pest-protection can also reduce the time required for cultivar development in some crops. Release of conventionally bred varieties of winter wheat that have the eyespot-disease-resistance gene *Pch1* required 13 years from the initial crosses, mainly because of time-consuming selections of lines with acceptable agronomic and disease-resistance characters (Jones et al. 1995).

It is important to recognize that transgenic resistance programs do not displace traditional breeding because transgenes alone cannot currently provide the full spectrum of agronomic traits necessary in commercial varieties. Furthermore, use of transgenes for resistance does not circumvent the normal process of agronomic quality assurance and testing that occurs throughout a well-managed breeding program.

Genetically Engineered Pest-protection Based on Novel Genetic Resources

The most publicized examples of engineered resistance based on novel genetic resources involve use of *Bacillus thuringiensis* (Bt) delta endotoxins (Estruch et al. 1997). Specific Bt endotoxin proteins are toxic to lepidopteran or coleopteran insects—many of which are destructive plant pests (such as the corn earworm and the tobacco budworm on cotton). Bt proteins in fermentation mixtures and spores have been used for decades in microbial formulations and by fermentation of *B. thuringiensis* strains that produce Bt crystalline proteins. Commercial transgenic varieties of corn, cotton, and potato that express Bt protein have been successful in reducing the incidence of pest damage and in reducing use of chemical pesticides in many cases (Robinson 1998; USDA 1999d; Gianessi 1999; Mullins and Mills 1999). These varieties may also be less susceptible to opportunistic pathogens that invade through wounds. The incidence of *Fusarium* ear rot and stalk rots in corn caused by several fungi may be significantly lower in Bt plants (Munkvold 1998). This would have the added benefit of lowering the exposure of humans and animals to fungal mycotoxins.

Pathogen-derived resistance involves the use of genes from a known pathogen in ways that result in protection from that pathogen (Beachy 1997; Sanford and Johnston 1985). The resistance can occur through a number of mechanisms. Expression of a normal or altered form of a pathogen protein in transgenic plants can disrupt the pathogen's normal pattern or timing of expression of that protein, or interfere with the interaction between a host and the pathogen. Coat protein-mediated resistance to viruses (Baulcombe 1996; Lomonossoff 1995) is the best-known example of pathogen-derived resistance and has been developed commercially in a number of crops. Expression of viral coat protein in plants interferes with uncoating of the viral genome and thereby prevents or delays the establishment of infection. Expression of multiple coat-protein genes confers resistance to multiple viruses (Tricoli et al. 1995). Expression of other types of viral genes that code for replicases and other proteins required for virus replication or movement in plants, has also been demonstrated to confer resistance in many cases (Baulcombe 1996; Lomonossoff 1995).

Pathogen-derived resistance can also trigger mechanisms that initiate or intensify natural plant-protection processes. For example, introduction of functional or nonfunctional viral transgenes into a plant often results in activation PTGS that suppresses expression of the transgene (Baulcombe 1996). The PTGS mechanism involves sequence-specific recognition and degradation of RNA in the cytoplasm (Grant 1999). Plants that activate PTGS to suppress a transgene invariably are highly resistant or immune to infection by the virus in which the transgene originated. In fact, PTGS of transgenes closely resembles the natural silencing response of plants to viruses, which can result in a recovery from the initial symptoms of infection (Al-Kaff et al. 1998; Ratcliff et al. 1997).

Genetically Engineered Resistance Based on Transfer of Natural Pest-Protection Mechanisms

The isolation of natural plant R and defensive genes provides the resources to transfer resistance from one plant species to another. Many of the known R genes, particularly those which confer protection from pathogen and nematode pests, are highly conserved in structure and function (Baker et al. 1997); that is, an R gene from one plant species will often function after transfer to another plant species. The N gene for resistance to tobacco mosaic virus in *Nicotiana tabacum*, for example, functions well after transfer to tomato (Whitham et al. 1996), and the *Cf-9* gene for race-specific protection of tomato from the fungus *Cladosporium fulvum* is functional when transferred to tobacco and potato, as the gene triggers HR specifically in response to the *C. fulvum avr9* avirulence protein (Hammond-Kosack et al. 1998).

Cloned R genes and pathogen avirulence genes make it possible to engineer natural resistance responses to a wide array of pathogens and pests. For example, combining an R gene with a corresponding avirulence gene under the control of appropriate regulatory genetic elements in transgenic pest-protected plants can facilitate activation of defense responses against pathogens that are normally not limited by that particular R gene.

Transfer of defense genes for specific degradative enzymes and inhibitors can also confer pest-protection. For example, constitutive or localized expression of a variety of genes that encode proteinase inhibitors, chitinases, and lectins in transgenic plants can provide protection against some chewing insects, sucking insects, or nematodes (Johnson et al. 1989; Kramer and Muthukrishnan 1997; Rao et al. 1998; Ryan 1990; Urwin et al. 1997). Transgenic modification of the production of defensive chemicals also will affect resistance to pests and pathogens (for example, Melanson et al. 1997).

Future Directions

Research focused on developing new ways to produce both conventional and transgenic pest-protected plants, is some of the most exciting in the field of plant biology. Through wide crosses and other nontransgenic techniques, plant resistance genes will continue to be transferred to crop species from species at greater and greater taxonomic distances. A number of genomics projects with model and crop plants are yielding data from which information about new R and defense genes can be obtained. That information could lead to identification and manipulation of resistance factors with unique specificities against important pests and pathogens. The signaling mechanisms whereby resistance responses are triggered by insects and pathogens are being unraveled. It might soon be possible to engineer plants with altered signaling components that result in resistance being triggered by a broader array of pests. Understanding how defensive secondary compounds and defense proteins are produced and regulated should allow design of resistant plants in which the active compound is expressed in defined locations within the plant. New Bt endotoxins with different specificities or targets are on the horizon, as are a variety of novel or pathogen-derived resistance strategies that have high efficiency and specificity.

A major goal of future development of pest-protected plants should be to decrease the potential for ecological and health risks that may be posed by some types of pest-protected plants (section 2.2.1). This work could include using marker-assisted breeding to avoid selecting varieties with inadvertently high levels of potential toxins and limiting expression of transgenes that have potential adverse nontarget effects to nonedible plant tissues. Development of strategies that enhance the effective life span, or durability, of transgenic pest-protection mechanisms is also of vital importance.

2.4 POTENTIAL HEALTH EFFECTS OF DIVERSE GENE PRODUCTS AND BREEDING METHODS

Sections 2.1 and 2.2 discussed standard risk-assessment terminology for GMPP plants and the 1987 NAS principles. One of the conclusions from those sections was that quantitative risk would vary case by case and depend on the gene product (hazard), its potency (hazard severity and dose-response relationship), and magnitude of exposure levels (exposure assessment). The following two subsections discuss the potential of various gene products (discussed in section 2.4.1) to cause adverse health effects and the potential of various genetic modification techniques (discussed in section 2.4.2) to cause indirect effects regardless of the intended gene product.

2.4.1 Health Effects Associated with Different Types of Gene Products

Different types of transgenic pest-protected plants that might be developed have the potential to cause adverse health effects. The degree of risk is related to the chance that potentially hazardous toxic or allergenic compounds are produced and to the magnitude of exposure of such compounds. The chance that hazardous compounds will be produced by either intended or unintended modifications is related largely to the specific type of transgene used.

Race-specific and Other Naturally Occurring Pest-Protective Genes

Pathogen race-specific pest-protective genes

Plants contain hundreds of pathogen race-specific pest-protective genes that are often referred to as R genes (see section 2.3). Most of these genes are evolutionarily conserved in structure and, most likely, in function. R genes for protection against pathogens are routinely transferred between plants by conventional breeding. There are no known toxic or nontarget effects of R gene products aside from their role in triggering localized and systemic defense responses in the presence of specific pathogens. Transfer of race-specific R-gene from a nonedible plant species to an edible species would result in new exposure of consumers and nontarget species to a specific R-gene product, although not to a new class of proteins. Because plants expressing R genes are not thought to mount a defense response unless encountered by pathogens, the chances of unintended health effects due to pleiotropic effects are remote. Compared with R-gene transfer by conventional breeding, introduction of an R gene via a transgenic method should result in fewer unintended effects, because of the lack of introduction of non-R gene DNA into the new variety.

Other pest-protective genes

Familiarity with the structure and function of race-nonspecific and other types of pathogen-protective genes is less than that with the major class of race-specific R genes, although their history of use in conventional breeding suggests that few hazards exist. Most of these pathogen-protective genes are probably conserved among different species of plants. It is unlikely that exposures to new classes of genes or gene products will result from transfer of these genes between plants, particularly if the source is an edible plant. However a protective gene may encode a protein that increases the concentration of one or more plant compounds with potential nontarget toxicity, thus leading to a potential hazard.

Defense Genes Encoding Pest-Degradation or Inhibitor Functions

A number of genes that code for degradative or hydrolytic functions, such as glucanases and chitinases, with pest-protective activities can be induced by infection (section 2.3.1). These types of enzymes are ubiquitous in plants and are common constituents of foods. Transfer of such genes from one edible plant to another is unlikely to cause a problem. However, if these genes were expressed at relatively high constitutive levels, were derived from nonedible plants or nonplant sources, or were engineered in such a way as to increase stability or alter structure substantially, novel exposures might result. The potential for adverse affects depends on the likelihood of increased toxicity or allergenicity of a novel product. The extent to which those properties are altered is partially predictable and testable.

Some inhibitor proteins, such as proteinase inhibitors, are effective defense proteins that are naturally induced by wounding or attack by chewing insects. Proteinase inhibitors are also present in relatively high concentrations in some food plants, such as potato. Animals can suffer adverse effects if foods high in proteinase inhibitors are consumed (Ryan 1990). Some naturally occurring lectins that have pest-protection attributes are also known to be toxic to humans and animals. Foods that are high in proteinase inhibitors and lectins, such as potatoes and beans, are usually cooked, and cooking destroys inhibitor and toxic activity. Depending on the protein, a plant modified to express high concentrations of inhibitors in edible tissues can cause adverse health effects if the plant is consumed raw, and such a risk can be reduced by designing transgenes that are expressed only in nonedible plant parts.

Genes Encoding Enzymes that Alter Secondary Metabolites or Natural Products

The wide variety of plant chemicals with potential pest-protection characteristics suggests that modification, transfer, or overexpression of genes that control natural-product biosynthesis can result in new types of pest-protected plants (sections 2.3.1 and 2.3.2). It is reasonable to predict that manipulation of those pathways can enhance resistance to insects and pathogens. The known toxicity of many protective natural products to nontarget organisms, however, means that such strategies could pose a risk. Furthermore, alteration of enzymes in one pathway might alter flux through other pathways. For example, Fray et al. (1995) demonstrated that constitutive overexpression of phytoene synthase, an enzyme required for carotenoid biosynthesis, in tomato had the unintended consequence of causing a dwarf phenotype, most likely due to decreases in gibberellin hormone and phytol (chlorophyll side chain) biosynthesis. Modulation of pathways for production of pest-protection chemicals could result in new exposures to potentially toxic compounds. That risk might be minimized by engineering transgenes with regulatory control elements that result in localized expression in nonedible tissues and plant parts. The risk might also be lowered through increased understanding of potentially hazardous compounds in commercial crop plants. Up-to-date and easily accessed databases with qualitative and quantitative descriptions of known or suspected toxicants would be particularly valuable in assessing inadvertent risks (sections 2.5.2 and 3.2.4) These databases could be used to catalog potential toxicants and their concentrations in edible portions of prominent cultivars grown under standardized conditions. New cultivars, regardless of how they were produced, could be tested for known or suspected toxicants and compared with established cultivars that are already being consumed.

Pathogen-Derived Protective Genes

Virus-derived transgenes

Because viruses of edible plants are common components of the food supply and no associations between such viral infections and adverse health effects have emerged, transgenic plants that express parts of viral genomes are generally considered not to represent an important human health risk because there is little chance of exposure to a novel virus gene product. The concentration of some viral gene products might be higher in a transgenic plant than in a naturally infected plant; but in the case of coat-protein-expressing plants, the concentrations will likely be lower.

Plants containing virus-derived transgenes that confer protection from pests because of activation of gene silencing generally produce very low concentrations of transgene-encoded protein, because the transgene mRNA is inactivated (section 2.3.2). From the standpoint of exposure to new or enhanced levels of viral protein, transgenic plants that contain silenced transgenes have little chance of causing problems.

Other pathogen-derived or pest-derived protective genes.

Experience with pathogen-derived or pest-derived protection from organisms other than viruses is sparse. It is difficult to assess this class of potential protective genes with regard to risk. New exposures could result, depending on the pathogen or pest, and potential toxicity to nontarget organisms is conceivable. Those examples will require relative-risk assessments case by case.

Genes from Sources Other than Plants, Plant Pathogens, or Pests

The various genes from novel genetic resources that confer pest-protection cannot be grouped from the standpoint of health risks. Transgenes that encode protective compounds from nonplant sources, such as Bt delta endotoxin, will probably present novel exposures and must be assessed on a case by case basis.

2.4.2   Indirect Effects Associated with Different Breeding Methods

To understand the risks posed by genetically modified pest-protected plants, it is important to understand that, in addition to the direct effects of the pest-protective gene (section 2.4.1), breeding can lead to indirect effects, such as the effects of extraneous genes linked to the protective genes and pleiotropic effects caused by the protective genes.

The potential for inadvertent changes caused by the addition of extraneous genes that are physically linked to protective genes depends on the breeding method used and the source of the protective gene. The breeding method and the source of genes used for breeding determine the amount of new DNA moved into the cultivar and the number of novel genes linked to the pest-protective gene. Therefore, genetic modification methods, both conventional and transgenic, are discussed below with regard to their potential for adding novel extraneous genes and their potential for causing unanticipated pleiotropic effects.

Conventional Breeding Methods that Involve Sexual Hybridization

The choice of parents used in the crosses and the mating structure of the plant species are important in determining the potential for inadvertent health effects associated with the progeny (hybrid, inbred line, or population).

Intraspecific hybrids of two cultivars

In crosses, or sexual hybrids, the amount of DNA transferred to the progeny can be immense. Depending on the mating design, a parent's contribution can range from very small (less than 1% for the donor parent in backcrossing) to very large (over 99% for the recurrent parent in backcrossing). In a cross between two parents that have been previously cultivated (that is, cultivars), each parent contributes one half its DNA to the progeny. In bread wheat, each parent contributes 16 billion base pairs to the progeny (see chapter 1, table 1.1). To put that into perspective, each wheat parent contributes roughly 55 times the total amount of DNA found in *Arabidopsis thaliana*. At first glance, the potential for inadvertent changes that could create new allergens or toxic compounds might seem high. However, through the long history of wheat improvement, wheat has remained a staple food consumed safely by much of the world (people suffering from celiac disease constitute a notable exception). The same is true of most major food crops.

In most conventional plant breeding, the goal is to create new genotypes that combine the favorable alleles from two cultivars into a superior progeny, which is selected. For cultivated crops without a history of alterations in antinutritional, allergenic, or toxic properties, the creation of new antinutritional, allergenic, or toxic properties due to crosses between two existing cultivars would be extremely rare.

Intraspecific hybrids of a cultivar and a wild relative, or interspecific hybrids (crosses between species in the same genus)

Within a species that includes a cultivar, there might be weedy relatives that can be used as a source of pest-protective genes (section 2.7.2). The potential for unexpected health effects in crosses between a cultivar and a weedy relative in the same species is similar to the potential for health effects associated with interspecific crosses described below.

Species in the same genus have a common ancestry and have numerous related or similar genes. Interspecific crosses are usually between a cultivar and a wild species that has a pest-protective gene of interest. Inasmuch as most of the wild relative's genes are removed by backcrossing, with the exception of those genes linked to the selected protective gene (known as a linkage block), the creation of a new antinutritional, allergenic or toxic constituent in the hybrid cultivar will be rare.

The size of the linkage block associated with the protective gene depends on the rate and type of recombination between the wild relative's chromosomes and those of the cultivated parent and on the number of backcrosses. Formulas used to estimate the size of the linkage block when there is normal recombination can be found in many papers (for example, Hanson 1959a,b; Muehlbauer et al. 1988). The practical outcome of moving a block of linked genes with the protective gene is that additional genes are carried with the protective gene (for example, Zeven et al. 1983). If homologous recombination is decreased because of poor chromosome pairing, the linkage block will be larger. Poor chromosome pairing occurs in interspecific and, more commonly, in intergeneric crosses, which are discussed below. However, as just mentioned, the species have a common ancestry and similar genes, so most of the genes in the linkage block will be related to those of the cultivar.

Intergeneric hybrids (crosses between genotypes of two genera)

In this case, there is greater divergence in the genetic ancestry, there are more genes that have never previously been combined, and the usually poorer chromosome pairing leads to larger linkage blocks. The genetic divergence is very important in the outcomes of these crosses. For example, bread wheat and rye (*Secale cereale* L.) are in different genera, but they share a relatively recent progenitor in evolutionary time. Hence, as in interspecific crosses, bread wheat and rye have many similar genes. The similarity in gene function between the two genera is illustrated by the great success of many bread-wheat cultivars that contain whole chromosome arms from rye that replaced the equivalent chromosome arms of wheat (Zeller and Hsam 1983; Lukaszewski 1990). The ability to replace a chromosome arm without decreasing productivity indicates that many of the genes in the two genera are equivalent.

The amount of DNA contained on an average chromosome arm for wheat would be about 380 million base pairs (16 billion base pairs in the haploid genome divided by 42 chromosome arms). In this example, short chromosome arms were replaced, so they would have fewer than the estimated number of DNA base pairs in an average chromosome arm. However, each chromosome arm would be as large as or larger than the *Arabidopsis* haploid genome. Historically, intergeneric crosses are usually used with other breeding methods (such as backcrossing) that remove most of one genus's genes from the commercial product. However, because of poor chromosome pairing, there is often little homologous recombination, and large linkage blocks of DNA are retained in the progeny (for example, the whole chromosome arms mentioned above). However, alien introgressions[3] into food plants have been common (for example, Friebe et al. 1996) and have been associated with little history of inadvertent problems. The example given above emphasizes rye introgressions into wheat where both plants are food crops, but many gene blocks introgressed into wheat are from grassy relatives whose grain is not consumed by humans.

New methods are constantly being developed to overcome interspecific and intergeneric hybridization barriers. As these barriers are overcome, the overall genetic difference between parents becomes larger. For example, it is now possible to create oat lines that contain corn genes through oat-corn hybridization. If an oat line with a gene for an allergen from corn were released, an oat consumer with corn allergy might no longer know which oat products were safe to eat.

Conventional and Transgenic Genetic Modification Methods that Do Not Involve Sexual Hybridization

Mutagenesis

Mutations include chromosome rearrangements (such as translocations, deletions, and transposable elements) and DNA changes (such as

---

[3]Integration of new blocks of DNA or new traits not previously found in the species.

single base changes, insertions and deletions). The typical variation we see in the traits of organisms generally involves naturally occurring mutations. Natural mutations and mutations induced by chemicals and radiation have been used to produce many commonly used cultivars. In general, few genes are modified and most of the genome is unaffected. Most mutations are from an active form to a less active form and would pose a problem only when the active form is needed to remove an antinutritional, allergenic, or toxic constituent.

Somaclonal variation

Somaclonal variation is considered to be a form of induced mutagenesis occurring during the tissue culture process. The reason for interest in somaclonal variation is that it increases the genetic variation in plants regenerated from tissue culture; one of the general procedures used to develop transgenic plants. Its potential for unfamiliar health effects would be similar to that of mutagenesis.

Somatic-cell fusion

Somatic-cell fusion has the potential to combine whole genomes from genotypes of widely divergent genera. However, it is rare for a somatic-fusion hybrid from widely divergent genera to be directly commercialized. More likely, the genes contributed from one genus would be reduced with another breeding technique or genetic manipulation (such as backcrossing). According to the rationale described above for sexual hybridization, somatic-cell fusion involving cells from the same species (intraspecific hybrids) would have a lower potential for adverse health effects than interspecific hybrids, which would have a lower potential than intergeneric hybrids.

Transgenic methods

Introduction of transgenes into plants typically involves random integration of DNA into the nuclear genome and the use of tissue culture, which can lead to somaclonal variation. If integration of a transgene occurs within or near a gene, there is a potential for unintended consequences. Disruption of a gene can lead to its down-regulation or inactivation. If the gene is essential, viable plants will not be recovered after the transformation or transgene introduction. If the gene is not essential for growth and development, viable transgenic plants will be recovered, but they might have unexpected traits.

Introduction of a transgene can also result in activation or up-regulation of an adjacent gene. In this case, the regulatory regions of the transgene stimulate a nearby resident gene, and potentially cause increased expression of that gene. It has been argued that one unintended consequence of this process is up-regulation of genes for biosynthesis of plant toxins. The potential for overproduction of hazardous compounds by this random integration of transgenes is likely to be similar to that for mutations, transposable-element activation, and other processes that yield genomic variation. The potential for adverse effects can be minimized through characterization of the locus of transgene insertion. Plants with transgenes that insert relatively close to genes known to affect production of potentially toxic compounds can be avoided.

It is important to point out, however, that these pleiotropic effects are not peculiar to transgenic plants. Crops resulting from conventional breeding and other nontransgenic methods can contain potentially hazardous concentrations of naturally occurring toxic compounds, as has been documented in new or established varieties. The introduction of whole chromosomes or chromosome arms from wild, nonedible relatives presents risks that are proportional to the number of alien genes added to the commercial cultivar. With sufficient testing, the risks posed by the introduction of inadvertent, potentially hazardous concentrations of known or suspected toxins into commercial transgenic or conventionally bred cultivars can be reduced.

2.5   POTENTIAL HUMAN HEALTH EFFECTS

In the United States, the EPA proposes to assess the health effects of pest-protected plants under the FIFRA and FFDCA (section 1.5). The FDA will regulate food safety and quality under FFDCA (P.L. 104-170). This section discusses EPA's scientific review of potential human health effects and general scientific issues surrounding those effects.

Although human health risks associated with transgenic pest-protected plants tend to be potential rather than apparent, some regard these potential risks as important and have expressed their views on appropriate testing and controls (OECD 1993a; Goldburg and Tjaden 1990). Potential food safety concerns for transgenic pest-protected plants (and other transgenic plant products) have been identified and articulated by EPA and FDA (FDA 1992) and international organizations (OECD 1993; FAO/WHO 1996; OECD 1997b). These key food safety considerations have served as a basis of the food safety reviews for the products currently in the market.

The potential risks of transgenic pest-protected plants to human health are generally related to the possibility of introducing new allergens or toxins into food-plant varieties, the possibility of introducing new allergens into pollen, or the possibility that previously unknown protein combinations now being produced in food plants will have unforeseen secondary or pleiotropic effects. The use of antibiotic-resistance marker genes has also given rise to concern[4].

In the regulation of recently approved transgenic pest-protected plant products (that is plant products with Bt and viral coat proteins), the emphasis has not been on detailed assessments of safety for humans or domestic animals. Rather, it has been on explaining the scientific basis for why there is probably no appreciable risk and justifying the tests which are required. Although the assumption of no appreciable risk from the recently reviewed transgenic pest-protected plants (for example, plants containing Cry1A and Cry3A Bt proteins and viral coat proteins) appears reasonable, it is important that the tests that are performed be rigorous, logical and scientifically sound. In most cases, the tests have these qualities. However, specific suggestions for improving the testing protocols can be found in sections 2.5.1, 2.5.2 and 3.1.3.

A number of traditional chemical pesticides are considered human carcinogens (Hodgson and Meyer 1997). Others have been linked to human health problems, such as Parkinson's disease (Fleming et al. 1994). Therefore,

Human health benefits could arise from reductions in the application of chemical pesticides due to the commercial production of certain transgenic pest-protected plants.

But it is not necessarily true that all traditional chemical pesticides pose a risk to human or domestic-animal health, so the benefits will depend on the particular pesticides that transgenic pest-protected plants replace and the effects of the transgenic pest-protected plant on pest control practices.

The proposed human health testing for EPA, and the testing for FDA consultation, fall into the categories outlined in box 2.2. Those categories are general, and considerable variation in methodology is possible. There is evidence that this variation has occurred under the current guidelines (see discussion below). Even though the EPA rule is not final, more specific testing protocols should be developed and adopted (see sections 4.3.5 and 4.3.7).

---

[4]Issues surrounding antibiotic resistance, although mentioned, were not analyzed in this report (section ES.1 and ES.2).

BOX 2.2
Categories of Human Health Testing for EPA and FDA

Health-effects assessment (general testing for potential hazards)
- Mammalian testing
- Digestibility assessment
- Allergenicity testing
- Homology with known food allergens and toxins

Human safety assessment (more specific assessments)
Food safety
- Compositional analysis
- Nutritional assessment (concentrations and effects on bioavailability)
- Unexpected or unanticipated effects
- Dietary exposure assessment
- Determination of substantial equivalence
- Animal-feed consideration

Nonfood safety (only EPA)
- Worker exposure
- Bystander exposure (for example, via pollen)

2.5.1   Toxicity and Allergenicity Tests During EPA Review

To provide a cross section of recent activity under the general guidance that EPA informally provides to prospective registrants, the committee examined EPA pesticide fact sheets and data provided by the registrant for three registered Bt toxins that are regulated as plant-pesticides: Bt Cry3A delta endotoxin in potato (EPA 1995a), Bt subs. *kurstaki* Cry1Ac delta endotoxin in cotton (EPA 1995b), and Bt Cry1Ab delta endotoxin in corn (EPA 1997a and 1998a). The committee also reviewed EPA pesticide fact sheets for Bt subs. *tolworthi* Cry 9C protein in corn (EPA 1998c).

In general, oral toxicity testing for Bt endotoxins is based on the presumption that there is unlikely to be a problem inasmuch as a number of Bt toxins have been widely used for many years in microbial sprays without human toxicity. A variety of Bt proteins have been subjected to toxicological testing, starting with testing conducted on microbial Bt products, which typically contain multiple Bt proteins. This testing included acute, subchronic and chronic toxicology testing of products during the 1960s and 1970s (EPA 1988b; McClintock et al. 1995). However, it should be noted that most previous field uses resulted in minimal toxin ingestion by humans because sprayed microbial Bt toxin only remains effective for an average of 1.5 days (largely because of ultraviolet degradation), and therefore, residues are rare on fruits and vegetables. Also, sprayed microbial Bt toxins are protoxins, while some Bt plants produce activated toxins.

Information in peer reviewed studies indicates that plant-expressed Bt proteins are probably without human health risk. Nevertheless, a minimal number of properly defined tests are needed to determine if based on plant modification of the proteins, or if based on use of more novel Bt toxins or Bt toxins not found in currently registered microbial products (for example Cry9C), there is a potential impact on human health. Post-transcriptional modification is known to occur in plants and such characteristics as the degree of glycosylation might also affect stability and other physiochemical properties of proteins. Tests should preferably be conducted with the protein as produced in the plant (see also section 3.1.3). However, the committee recognizes that it is often difficult to obtain enough plant-expressed protein; in these cases, the committee recommends that

The EPA should provide clear, scientifically justifiable criteria for establishing biochemical and functional equivalency when registrants request permission to test non plant-expressed proteins in lieu of plant-expressed proteins.

The strong likelihood that gene products currently found in commercial transgenic pest-protected plants are not allergens does not remove the need for a minimum of properly planned and executed tests. For example, allergenicity is assumed to be unimportant for many Bt endotoxins, more because of the common characteristics of food allergens than because of rigorous testing. The Cry 1Ab pesticide fact sheet (EPA 1998a) states that

> current scientific knowledge suggests that common food allergens tend to be resistant to degradation by heat, acid and proteases, are glycosylated and present at high concentrations in the food. The delta endotoxins are not present at high concentrations, are not resistant to degradation by heat, acid and proteases, and are apparently not glycosylated when produced in plants.

In the case of Cry3A in potatoes (EPA 1995a), the company demonstrated that the endotoxin is not a major component of the food, is apparently not glycosylated in plants, and is digested by gastric enzymes. However, Cry9C toxin, unlike the Cry1A and 3A toxins, does not degrade rapidly in gastric fluids and is relatively more heat-stable (EPA 1998c); these characteristics of Cry9C raise concerns of allergenicity. It is important to note that levels of gastric enzymes may vary among individuals and that those variations may need to be considered.

Although the standard tests indicate nonallergenicity for Cry3A, they were not all carried out on the endotoxin produced in potatoes, and none involved testing the immune system itself. Allergenicity is difficult to test, in part because prior exposure is a prerequisite to an allergic reaction, but tests for allergenicity ideally should involve the immune system or use an allergic endpoint. Useful guides to protein allergenicity include a supplement to Critical Reviews in Food Science and Nutrition (Metcalfe 1996a) and the proceedings of a workshop held at the 1998 Society of Toxicology annual meeting and recently published (Kimber et al. 1999). They make it clear that food allergy is relatively common and can have numerous clinical manifestations, some of which are serious and life-threatening. Furthermore, it is well established that allergenic proteins can be found in many food plants, of which some, such as soybeans and potatoes, have been genetically modified for pest-protection, and many others are or will be candidates for this type of genetic modification.

Those two sources also summarize the problems in protein allergenicity testing. Some, such as the double-blind placebo-controlled food challenge and the skin-prick test, although they involve adverse toxicologic endpoints and are carried out directly on human volunteers, nevertheless provide questionable results, in that they require volunteers who are already sensitized and that they are difficult to implement with novel proteins that have not traditionally been consumed. Tiered tests involving protein homology and stability comparisons with known food allergens and immunoassays for specific classes of antibodies are also proposed in these documents and are currently used by the agencies as a screen for allergenicity (figure 2.1). However, the tests in figure 2.1 either are indirect, do not involve adverse effects, or are otherwise problematic for testing of novel proteins that have not previously been components of the food supply. Indeed, figure 2.1 starts with a decision based on whether or not the protein is derived from a source that is known to be allergenic. This decision can usually be made clearly if the source is a food plant. For transgenic proteins such as Bt endotoxins making such an assessment would be complicated. If we conservatively choose the "yes" decision, then it would be extremely difficult to complete all of the tests listed because test materials and previously exposed human subjects are not readily available.

It is clear from the report of Nordlee et al. (1996) on the expression of Brazil nut protein in soybeans that genetic engineering can result in the expression of an allergenic protein in a food plant, but this is not known to be the case with any commercialized transgenic pest-protected plant. However, some testing of pest-protected plants and purified gene products is appropriate in many cases inasmuch as allergenicity is one of the possible adverse effects. The possibility that proteins in spore-crystal formulations of Bt can interact with the human immune system was suggested by a recent study on workers in crops topically treated with Bt

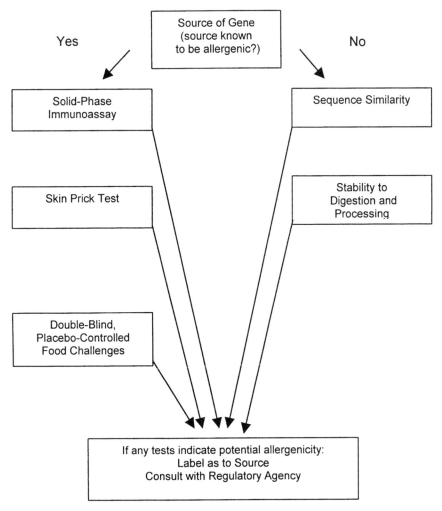

FIGURE 2.1 Tests for Potential Allergenicity. Source: Adapted from Metcalfe (1996b).

sprays (Bernstein et al. 1999). The response seen in this study could have been due to proteins other than the Bt toxins in the formulation, so similar studies should be performed with purified Bt toxins.

2.5.2 Questions Concerning Health Impacts

The potential for transgenic pest-protected plants to pose a threat to human or animal health must be considered against the background of existing information. To date no such effects have been shown with commercialized transgenic crop plants. The work of Ewen and Pusztai (1999) hints of some possible interaction between a lectin expressed in potato and alterations in the potato caused by the genetic engineering process. According to the study, diets containing genetically engineered potatoes expressing the lectin, *Galanthus nivalis* agglutinin (GNA), showed some effects on different parts of the rat gastrointestinal tract. Those effects fell into two categories, ones caused by the GNA transgene itself and others caused by pleiotropic effects of expressing the transgene. However, analysis of the work of Ewen and Pusztai by the Royal Society (Royal Society 1999) and by Kuiper et al. (1999) indicates that the study lacked scientific rigor. For example, data concerning the biochemical composition of the potatoes used in the study show that the nontransgenic variety differed significantly from the transgenic variety. These differences could be attributable to natural variations in potato lines and are not necessarily due to the genetic modification (Kuiper et al. 1999).

It is important to ask whether any such threats have resulted from more conventional genetic modification of agricultural crop plants (conventional pest-protected plants) and, if so, whether they can serve as examples for assessing the risks of transgenic pest-protected plants. The questions outlined below can be asked, with equal validity, of transgenic pest-protected or conventional pest-protected plants. It should be noted, however, that conventional breeding usually selects for endpoints that are almost always controlled by several genes. It could be reasoned that such selection because of genetic linkage would be more likely to select simultaneously and inadvertently for an additional undesirable characteristic than would the introduction of a single gene or even a small number of genes with transgenic techniques (section 2.4.2). Both animal and plant breeding have yielded examples of inadvertent selection of undesirable characteristics.

Is the Transgene Expressed in the Edible Part of the Plant?

If so, what are the potential effects on humans, domestic animals, and other nontarget animals? Toxicity testing can be carried out on the gene product or the edible part of the plant, and testing to predict potential effects on humans is carried out on laboratory animals and extrapolated to humans. Toxicity testing of chemicals that are macronutrients, such as proteins, has unique problems. The maximal tolerated dose, as determined in short term tests, is usually very high and palatability problems unrelated to toxicity can interfere with tests. Given the high dose, it is difficult to feed enough of the plant material to the test subjects without making substantial dietary changes. Suitable controls are difficult to devise because the control diet should have the same nutritional characteristics as the control diet.

It has been suggested (for example, Health Canada 1994) that in cases of foods where the concentration of substances to be tested cannot be increased, it would be useful to test the plant material in longterm feeding experiments with animals whose natural diets consist of large quantities and the type of plant material being tested. Research on longterm feeding of transgenic pest-protected plants to those animals (for example, grain and forage crops to livestock) might provide information relevant to human health effects (see also section 3.4). Such testing has been shown to be practical with herbicide-tolerant soybean (Hammond et al. 1996), and with Bt corn (Jackson et al. 1995). Livestock that are normally fed on the crop in question can be fed on the genetically altered variety from weaning until a termination time consistent with normal agricultural practice. The genetically closest variety can be used as a control. These types of studies would make use of the natural diet of the test organism to test large quantities of the whole plant. Nonmodified plant varieties that were identical to the genetically modified plant before the modification occurred should be used as controls. Most toxicity testing is conducted using the purified plant-pesticide (section 3.1.3), and therefore pleiotropic effects of the genetic modification cannot be monitored. If proper controls are used, feeding whole plants to the test animals might allow for the detection of potential toxicity due to pleiotropic effects.

However, there will be a need to carefully assess the relevance of such testing to human health. Humans and ruminants have different digestive systems, as humans are mongastric and ruminants have a four-chambered stomach that can serve as a buffer from the effects of some proteins. Feeding studies using monogastric animals, such as hogs, whose natural diets consist of the transgenic crop in question may provide more relevant information. While the finding of negative effects in such livestock tests would certainly raise concerns, the finding of no effects on these animals is hard to interpret because we lack sufficient information on their biochemical similarities to humans.

Is the Physiology of the Plant Changed by the Genetic Modification?

Potential pleiotropic effects of genetic modification on plant physiology and biochemistry are discussed below. The committee concludes that

It is important to monitor for physiological and biochemical changes during the development of transgenic pest-protected plants.

However, there is lack of data on the natural levels of endogenous plant compounds in both transgenic and conventional pest-protected plants and on how these levels vary with the plant's environment (see sections 2.4.1, 3.2.4, and 3.4.1).

In addition, any changes in the use of traditional pesticides may be considered in parallel to these pleiotropic effects, as the benefits of decreased use of toxic pesticides may offset the risks of increased toxicity due to genetic modification.

Changes in the levels of toxic secondary plant chemicals

Secondary plant chemicals (chemical compounds that are not required for normal growth and development of the plant) can be toxic to humans and other mammals (Senti and Rizek 1974), and the concentrations of these chemicals can be changed, either purposely or inadvertently, by conventional or transgenic genetic modifications. For example, potatoes, a major source of starch and good-quality protein, contain toxic glycoalkaloids to which humans appear to be more sensitive than other mammals. The modes of action seem to be cholinesterase inhibition in the nervous system and disruption of cell membranes in other organ systems (Friedman and McDonald 1997). This is similar to the mode of action of organophosphate and carbamate insecticides. Many factors, both genetic and environmental, affect glycoalkaloid concentrations in the potato and distribution to different plant parts.

A number of confirmed cases of human poisoning have been caused by potatoes with high glycoalkaloid concentrations (Friedman and McDonald 1997). The conventionally-bred Lenape variety of potato is an example of why great caution must be exercised in the development of new varieties of food plants that contain human toxins. The Lenape variety was developed by crossing *Solanum tuberosum* and *S. chacoense* to incorporate resistance to certain pests based on leptines (Sturckow and Low 1961). This variety was released for commercial use in 1967 (Akeley et al. 1968), but was soon withdrawn from the market (Sinden and Webb 1972). The new variety was indeed resistant to pests and had other desirable characteristics, but there were reports of illness, caused by ingesting tubers with high glycoalkaloid content (Zitnak and Johnson 1970). In Sweden, a popular commercial potato variety, Magnum Bonum, was withdrawn from the market for similar reasons (Hellenas et al. 1995).

Another problem of potential importance, the appearance of toxins that were not present in the parental lines, also has been demonstrated in potato. Somatic hybrids between *Solanum brevidens* and *S. tuberosum* contained the steroidal alkaloid demissine, not found in either parental line. Laurila et al. (1996) advanced the plausible hypothesis that the hydrogenase enzyme of *S. brevidens* that produces tomatidine from the precursor teinamine by hydrogenation of the double bond at position 5 also produces demissidine by hydrogenating the corresponding double bond in solanidine, a compound that is found in *S. tuberosum* but not in *S. brevidens*.

Celery is another example of where conventional plant breeding methods produced an unwanted result. Furanocoumarins are bioactive components of celery and are known to interact with DNA, are mutagenic and carcinogenic, are reproductive toxicants and cause contact dermatitis in humans. A celery line with resistance to *Fusarium,* but with concentrations of linear furanocoumarins high enough to cause severe contact dermatitis in field workers, was almost released for commercialization (Diawara and Trumble 1997; Trumble et al. 1990 and in press).

Transgenic pest-protected varieties commercialized to date in the United States have not been shown to have elevated levels of certain toxic compounds. For discussion during FDA consultation, companies monitor for changes in the levels of certain endogenous plant compounds. For example, ringspot-resistant transgenic pest-protected papaya lines were shown by chemical analysis to have no more of the suspected human toxicant benzyl isothiocyanate than nontransgenic varieties (University of Hawaii 1997).

Changes in the distribution of secondary plant chemicals so that they are expressed in edible parts of the plant

The "edible" part of a plant varies with the species and the consumer in question. In the human diet, the part eaten can also vary with the cultural background of the consumer. Overall increases in the concentrations of secondary plant chemicals in the total plant might cause toxic chemicals that are normally present only in trace amounts in edible parts to be increased to the point where they pose a toxic hazard. In some cases, genes transferred by conventional breeding can also change the distribution of secondary plant compounds among plant parts. For example, cucumber was bred to produce a bitter protective compound, cucurbitacin, in the leaves and stems of the plant but not in the fruits. A single gene controls the restriction of the compound to the leaves and stems (Barham 1953).

Changes in the chemical or physical properties of the plant surface or edible parts in ways that affect its contact allergenicity or food allergenicity, respectively

Some plants cause either contact allergic responses or food allergies. However, only a few documented examples involving contact allergens

(Trumble et al. 1990 and in press) show that allergens can be inadvertently changed during conventional breeding, and there is no substantial body of information on this possibility. There are some examples of assessments of endogenous allergens in transgenic plants (Burks and Fuchs 1995; Metcalfe et al. 1996a).

2.5.3   Summary

The above questions can be used to guide the review process for the potential health effects of transgenic pest-protected plants. In reviewing toxicity testing relevant to human health for currently commercialized transgenic pest-protected plant products (that is, Bt toxins and viral proteins), the committee found that,

When the active ingredient is a protein, short-term oral toxicity and potential allergenicity testing are currently appropriate, inasmuch as the testing protocols are the ones currently available. However, testing protocols, particularly for allergenicity, should be improved with additional research.

The committee recommends that

When the active ingredient of a transgenic pest-protected plant is a protein and when health effects data are required, both short term oral toxicity and potential for allergenicity should be tested. Additional categories of helath effects testing (such as for carcinogenicity) should not be required unless justified.

Additional categories of toxicity testing do not appear justified for currently commercialized products (such as products with Cry1A and 3A Bt endotoxins and viral coat proteins). However, in these cases, it important that the tests that are performed be rigorous, logical, and scientifically sound (see sections 2.5.1, 2.5.2, and 3.1.3). Novel, or less familiar plant-pesticides (that is, in comparison to viral coat proteins and Bt toxins) may require additional toxicity testing.

In addition, given the difficulties with determining the potential allergenicity of proteins not currently in the food supply, the committee recommends that

Priority should be given to the development of improved methods for identifying potential allergens in pest-protected plants, specifically, the development of tests with human immune-system endpoints and of more reliable animal models.

Protocols for toxicity- and allergenicity-testing, whether of gene products or plants, should be reconciled across agencies (see section 4.3). Variations among EPA, FDA, and USDA toxicity-testing protocols, other than those dictated by legislative authority, should be minimized or abolished. It should also be noted that the use of "familiarity" as a guideline to minimize testing can sometimes be inappropriate and warrants caution. Familiarity can be defined as indirect knowledge or experience obtained from similar gene products, plant varieties, or progenitor varieties grown under similar conditions and used for the same purposes in the same way.

2.6 POTENTIAL EFFECTS ON NONTARGET ORGANISMS

In addition to human health concerns, there is concern that gene products of some transgenic and conventional plants may be toxic to nontarget species in the ecosystem. This section reviews and discusses relevant data on such potential nontarget effects of transgenic and conventional pest-protected plants.

2.6.1 Definitions

It is useful to divide the effects of conventional and transgenic pest-protected plants on nontarget species into direct and indirect effects. Direct effects include the adverse effects of the toxic plant components on nontarget herbivores, omnivores, and microorganisms that feed on live plants (Hare 1992) and on detritivores that feed on dead plants (Horner et al. 1988). When toxic substances are on the surface of a plant, there could also be direct effects on organisms that do not feed on the plant solely caused by contact toxicity or repellence (Farrar and Kennedy 1993).

Indirect effects include adverse effects of genetically modified pest-protected (GMPP) plants through an intermediary species (Bergman and Tingey 1979; Hare 1992; Price et al. 1980). For example, if an herbivore is tolerant of toxic substances in a plant, these substances could be found in the herbivore's digestive system (Price et al. 1980) or even be sequestered in the herbivore's tissues (Brower et al. 1984; Duffey 1980; Tallamy et al. 1998). Such a herbivore can be unpalatable or toxic to some of its predators or parasites because of this biological magnification of plant-defense compounds (Brower et al. 1984; Ferguson and Metcalf 1985; Hoy et al. 1998). If a pest-protected plant causes dramatic decreases in some herbivore or omnivore populations, there will be less nutrient material for the next level in the food chain. It is theoretically possible that a specialized predator, parasite, or pathogen of an affected herbivore could become locally extinct (Riggin-Bucci and Gould 1997).

The few data sets on nontarget impacts of transgenic pest-protected plants come mostly from crops that express Bt toxins (Hoy et al. 1998). However, a considerable number of studies have examined direct and indirect effects of conventional pest-protected crops and of wild host plants that differ in their pesticidal properties. Some of these studies are reviewed in the following sections (2.6.2 and 2.6.3).

2.6.2   Direct Effects

Effects of Structural Changes

A number of studies have shown that the leaf hairs and leaf hair exudates found in resistant cultivars can kill predators and parasitoids of insect pests directly (Bottrell and Barbosa 1998; Price et al. 1980). This can result in a loss of biocontrol of the target pest. A classic study by Rabb and Bradley (1968) demonstrated that the sticky trichomes on tobacco leaves significantly decreased parasitism of the tobacco budworm (*Heliothis virescens* F.). A study by Kauffman and Kennedy (1989) demonstrated that allelochemicals in the trichomes of resistant tomatoes were toxic to a parasitoid (*Campoletis sonorensis*) of the corn earworm (*Helicoverpa zea*). The trichomes in these tomato lines also had an adverse effect on other parasites and predators (for example, Farrar and Kennedy 1993; Kashyap et al. 1991). A study of trichomes of potato plants showed that the impact of increased trichome density on natural enemies seen in the greenhouse was decreased under some field conditions (Obrycki and Tauber 1984); the study underscored the importance of creating experimental conditions similar to those experienced by insects in typical agroecosystems.

Changes in glossiness of leaves (for example, Eigenbrode et al. 1995) and general plant architecture (e.g., Kareiva and Sahakian 1990) can also affect the efficacy of natural enemies. Because plant-produced volatile substances attract many natural enemies, a new cultivar with an altered profile of insecticidal volatile chemicals could have reduced attraction to parasites or predators resulting in decreased biological control of the pest (Bottrell and Barbosa 1998; Dicke 1996).

Effects of Plant Ingestion

The chemical composition of forage plants can have a significant effect on livestock and bees (Keeler et al. 1978). Some chemicals can be passed on in the milk of cows and goats (Dickson and King 1978) and the honey of bees (Bull et al. 1968). Plant breeders have long recognized that some cultivars of forage crops can adversely affect livestock growth and health (Reitz and Caldwell 1974). For example, reed canary grass varies in palatability as a forage crop: the least palatable cultivars have the highest alkaloid concentrations (Williams et al. 1971). Cultivars of alfalfa vary in content of saponin which can depress the growth of chickens (Hanson et al. 1973; Reitz and Caldwell 1974). The terpenoid gossypol makes cotton resistant to caterpillar pests (Shaver and Lukefahr 1969), but cottonseed meal from high gossypol cultivars is poisonous to swine and causes darkened yolks in eggs of chickens (Reitz and Caldwell 1974).

Some plant toxins similar to those in pest-protected plants are relatively stable and can be found in decaying plant tissues (Horner et al. 1988). Only a few studies have examined the effects of such compounds on detritivores (Horner et al. 1988; Paavolainen et al. 1998). One study of balsam poplar showed that the tannins and phenolic chemicals in fallen leaves could inhibit mineralization through two interactions with detritivores. The result can be reduced soil-nitrogen availability (Schimel et al. 1996).

Bt crops have not had an impact on honeybees in tests conducted for EPA approval of Bt crops (EPA 1998c). However, at high concentrations the Cry1Ab toxins engineered into crops have been shown to be toxic to *Collembola* that are part of the detritus food chain (EPA 1997a). Another Bt toxin, Cry9C, was not found to be toxic to the same *Collembola* species (EPA 1998c). Cry1Ab but not Cry9C was found to have some toxicity to *Daphnia* (EPA 1997a and 1998c).

Few studies have focused on measuring the impact of pest-protected plants on the population dynamics of insects (reviewed in Gould 1998). Likewise, few have examined the direct physiological effects of Bt on nontarget herbivores, but some nontarget lepidopterans feeding on crops that contain Cry1A Bt toxins are likely to be affected. These nontarget insects would include lepidopterans that are not pests (that is, do not provide significant damage to the crop) and lepidopteran pests that are not sufficiently affected enough by a pest-protected plant for pest-protection of this kind to be economically useful.

Effects of Pollen Ingestion by Nontarget Herbivores

Pollen from wind-pollinated pest-protected plants can be deposited on nearby vegetation (figure 2.2) and inadvertently ingested by nontarget leaf-eating insects. For example, a small fraction of corn pollen is known to disperse up to180 ft from the edge of the crop (Raynor et al. 1972) and can be deposited on milkweed (*Asclepias* sp.), which are common in and along edges of corn fields in the midwestern United States where about half the population of US monarchs spend some of the summer (Wassenaar and Hobson 1998). Milkweed is the only food of monarch

(a)

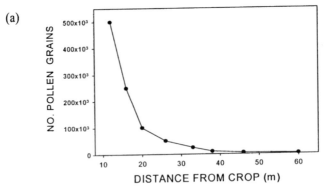

(a) Typical downwind deposition of corn pollen on greased microscope slides at ground level; measured after 9 h with no rain and light winds. Source: Raynor et al. (1972).

(b)

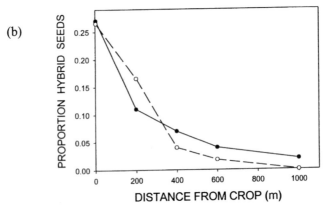

(b) Proportion of hybrid seeds of wild and crop sunflowers planted near each other; data from two sites. Source: Arias and Rieseberg (1994).

FIGURE 2.2 Examples of pollen and gene dispersal as a function of distance from edge of crop. Dispersal distances are expected to depend on local conditions and relative sizes of donor and recipient groups of plants.

butterfly larvae. Laboratory studies showed that young monarch larvae were killed by ingesting high doses of Bt corn pollen that had been experimentally dusted onto milkweed leaves (Losey et al. 1999). Another study by Hansen and Obrycki (1999c) obtained similar results when monarch larvae were placed on milkweed leaves that had been collected at the edges of Bt and non-Bt corn fields in Iowa and brought into the laboratory. Although both studies should be viewed as preliminary and actual negative impact, if any, on the population densities of monarch butterflies is yet unknown, they have focused attention on the question of whether widespread dispersal of insecticidal pollen can occur if entire regions are planted with wind-pollinated Bt plants (such as corn, poplar, and pine). It should be noted, however, that some Bt corn cultivars apparently do not produce any Bt in their pollen (EPA 1998b). In those cases, Bt toxin would not be a potential hazard to nontargets even if pollen dispersed great distances.

One recent field test indicates that at least 500 Bt pollen grains per square centimeter is necessary to sicken monarch caterpillars and that milkweed plants growing adjacent to corn fields had only an average of 78 grains per square centimeter (Kendall 1999). Eighty-eight percent of milkweed within one meter fell below the level of toxicity to caterpillars. On the other hand, laboratory tests by Hansen and Obrycki (1999b) demonstrated that when monarch caterpillars consumed milkweed leaves experimentally dusted with 135 grains per square centimeter of pollen (comparable to pollen concentrations they found in the field), there was 46% mortality if the Bt pollen source was insertion event Bt11 and 65% if the source was event 176. Moreover, research suggests that wind direction, rainfall, and other factors can significantly affect pollen concentration (Weiss 1999). Further field-based research is needed to determine whether dispersed Bt pollen could have detectable effects on the population dynamics of nontarget organisms.

2.6.3 Indirect Effects

Studies in which predators were fed insects that developed on pest-protected cultivars have often produced adverse effects on the predator. For example, the compound DIMBOA is found in lepidopteran-resistant grain crops, including corn. Ladybugs that consumed aphids that had fed on diets containing DIMBOA had slower development (Martos et al. 1992). The tobacco hornworm parasitoid *Cotesia congregata* is adversely affected when the hornworm feeds on tobacco or artificial diets with high nicotine concentrations (Thorpe and Barbosa 1986). Cucumbers that are resistant to spider mites typically have higher concentrations of the triterpenoid cucurbitacin-C than nonresistant cultivars (DaCosta and Jones 1971; Gould 1978). Spotted cucumber beetles that feed on the resistant cultivars sequester cucurbitacin-C in their bodies and eggs (Ferguson and Mecalf 1985). The cucurbitacin-C in their eggs keeps the eggs from being killed by the entomopathogenic fungus *Metarhizium anisopliae* (Tallamy et al. 1998).

Although a pest-protected plant could have adverse effects on beneficial insects in the agroecosystem, the value of the cultivar is generally based on the balance between adverse effects on beneficial and detrimental species. For example, Kauffman and Flanders (1985) found an adverse effect of pest-protected soybean on a parasitoid of the Mexican bean beetle, but the effect on the beetle pest was greater than on the parasitoid; it was therefore concluded that overall the cultivar would be useful. Hare (1992) reviewed the published data on overall effects of pest-protected host plants on biological control. Of the 16 case studies he reviewed, there were antagonistic interactions in six, positive synergistic interactions in two, neutral effects in five, and a concentration-sensitive relationship in three.

There have been a few published studies of the indirect effect of transgenic Bt-producing crops on natural enemies. Depending on the species involved , some studies reported no significant effects (for example, Hough-Goldstein and Keil 1991), but others have found adverse effects. Giroux et al. (1994) reported that the ladybug predator of the Colorado potato beetle consumed fewer potato beetle eggs when potato Bt levels were high. Hillbeck et al. (1998a, b) found that when chrysopid larvae were reared on prey that were fed Bt-producing corn, they suffered 62% mortality. When they were reared on prey that were fed non-Bt corn, mortality was only 37%. Those results were found with two prey species, one sensitive to Bt and the other insensitive. The results indicate that it was the Bt toxin in the bodies of the prey, and not simply unhealthy prey, that caused the heightened mortality.

It is important to ask whether such indirect effects will have a harmful effect on the agroecosystem. In some cases, the use of conventional pest-protected plants has lead to decreased foraging efficiency of predators and parasitoids (Boethel and Eikenbary 1986). A few field studies have attempted to measure the effect of transgenic Bt potatoes on the diversity of insects in crop fields (Hoy et al. 1998). In one 4-year study at five locations in the Wisconsin potato-production system, it was found that populations of predators were, on the average, 63.8% lower in fields where non-Bt potatoes were managed with typical insecticides than in fields of transgenic Bt potatoes. Parasitoid populations in the chemically treated potato fields were, on the average, 58.4% lower than in Bt potato fields. Insecticidal treatment of potatoes to control Colorado potato beetles with conventional insecticides often results in secondary outbreaks of aphids because the aphids are released from biological control. In three of the four years of the study, aphid populations (which are not affected by Bt toxins) were lower in the transgenic Bt potato fields than in chemically treated potato fields, presumably because of greater biological control.

A study of potatoes in Ohio reported similar results (Hoy et al. 1998). In potato fields treated with pyrethroid insecticide to manage Colorado potato beetles, natural enemies were reduced to undetectable levels. This resulted in aphid outbreaks that could be only marginally controlled by insecticides (253 aphids per meter of row in late July). In the transgenic Bt potatoes where natural enemies were always detectable (1 to 4 per meter), the aphid density was very low (below 10 per meter).

Those two studies compared the impacts of chemically-intensive potato farming and farming with transgenic Bt-potato. No comparison with a no-chemical control treatment was presented, but it might have been inappropriate. Another study provided to the committee indicated that there were no significant differences in beneficial arthropods between plots in which transgenic Bt potatoes or microbial Bt sprays were used (Feldman et al. 1994). Both had similar densities of beneficial arthropods, while plots in which chemical insecticides were used had much lower densities of some of these arthropods (Feldman et al. 1994).

In most of the areas where Bt potatoes will be used commercially, the crop is usually protected against the target Colorado potato beetle with conventional pesticides. In the corn system, most of the acreage has not been treated for the European corn borer, the major target pest of the Bt corn cultivars, because the insect feeds mostly within the plant stem where pesticides are typically not effective. In addition, pesticide treatment is more expensive than the yield losses due to the borer. In this case, the appropriate comparison of environmental impacts of the transgenic Bt cultivars would be with a system where non-Bt corn was not treated with pesticides aimed at the European corn borer.

It is useful to ask what will happen to the long-term biodiversity of agroecosytems if biotechnology provides us with crops that are constitutively toxic to almost all insect herbivores. A study of insecticide effects on quail populations in soybean fields provides interesting insight. Palmer et al. (1998) found no physiologic effects of the pesticide residues on the quail. However, quail in sprayed fields had lower weight gain and lower survival than those in control fields because there were fewer insects to feed on (Palmer 1995). If in the future we continue to commercialize pest-protected crops that constitutively express more diverse insect-specific toxins, we could inadvertently produce crops that lower general herbivore abundance. This could result in lower biodiversity of species at higher trophic levels that depend on herbivorous insects as food.

2.6.4 Summary

Conventional and transgenic pest-protected crops can adversely affect nontarget organisms through direct contact with or ingestion of the plant or pollen by the nontarget organisms and through indirect contact when the pest-protective substances (or their effects) are passed to other trophic levels that are consumed by nontarget organisms. However, the committee found that

Both conventional and transgenic pest-protected crops could have effects on nontarget species, but these potential impacts on nontarget organisms are generally expected to be smaller than the impacts of broad-spectrum synthetic insecticides, and therefore, the use of pest-protected plants could lead to greater biodiversity in agroecosystems where they replace the use of those insecticides.

Current criteria for commercialization of transgenic pest-protected plants includes several laboratory toxicity tests for nontarget organisms (see appendix B and section 3.1.2). In light of the above discussions, more field evaluations should be conducted to determine the impacts of specific pest-protected crops on nontarget organisms, compared to impacts of standard and alternative agricultural practices. The committee recommends that

Criteria for evaluating the merit of commercializing a new transgenic pest-protected plant should include the anticipated impacts on nontarget organisms compared with those of currently used[5] pest control techniques.

2.7   GENE FLOW FROM TRANSGENIC PEST-PROTECTED PLANTS

Genes from one crop plant may be spread to other plants of the same or related species when pollen is transported by wind, bees, or other animal pollinators. Genes have been flowing from crops to weedy relatives of crop plants for centuries. Now it is also possible for fitness-enhancing transgenes to spread to weed populations. In this report, the committee uses the terms "weedy" and "invasive" in reference to plants that are unwanted in human-dominated or natural habitats. Many people think of weeds primarily as undesirable plants that infest agricultural fields, tree plantations, lawns, and other managed areas. However, natural and semi-natural habitats such as wetlands, coastal dunes, and rangelands are also harmed by the spread of weedy species. Weedy plants quickly colonize open space and may displace non-weedy species, as has occurred with kudzu, Scotch broom, spotted knapweed, and purple loosestrife, for example. Once established, these types of plants are often difficult to eradicate. Annual species release large quantities of long-

---

[5]Includes both chemical and non-chemical methods which are currently used.

lived seeds that persist in the soil, while perennial species can be difficult to kill with herbicides, especially when they occur outside the realm of farmers' fields. Although only a small percentage of the weedy and invasive species in the United States are capable of crossing with cultivated plants, these species merit special attention with regard to crop-to-wild gene flow (NRC 1989). In addition, species that are not currently designated as weeds could potentially become more difficult to control if their populations are released from key ecological constraints (for example, herbivory, disease, or drought stress). This process is discussed further in the 1989 NRC book on field testing genetically engineered organisms.

The 1989 NRC report noted that "the potential for enhanced weediness is the major environmental risk perceived for introductions of genetically modified plants" (NRC 1989, p. 3). To evaluate problems possibly associated with gene flow from transgenic pest-protected plants, it would be useful to know how much pollen is dispersed from a crop, how far it travels, and whether its genes persist in wild populations. It would also be useful to know whether the spread of transgenes via pollen can lead to increased invasiveness in wild relatives of the crop.

Misinformation about these issues causes people to regard gene flow itself as a threat to the environment. Gene flow is indeed common, but the process of gene flow is important only if it leads to undesirable consequences, such as the spread of transgenes that are advantageous to weeds. In other words, the *process* of gene flow does not pose a hazard itself, but the *consequences* of gene flow might. Gene flow is the *process* by which exposure can occur (section 2.1). The *consequences* or potential hazards are dependent on the nature of the transferred trait, its level of expression in the recipient plant, and the biology and ecology of the recipient plant. The fact that the plant is genetically modified generally does not affect the *process* of gene flow or pollen dispersal.

In this section, the committee highlights the scientific dimensions of gene flow and its potential consequences. A few examples of transgenic pest-protected plants from which the consequences of gene flow are serious enough to merit special attention with regard to the weediness of wild relatives are also highlighted. The committee also describes the extent of gene flow from transgenic crops to organically grown crops of the same species. Scientific guidance as it relates to evaluating the consequence of gene flow and the review of gene flow for environmental safety assessments are presented in sections 3.1.4 and 3.3.

2.7.1 Pollen Dispersal and Outcrossing

The distance that pollen moves varies widely among species. Some crops—such as rice, wheat, and soybean—are mostly self-pollinated (selfed). In sorghum, most seeds result from self-pollination, and the rest

TABLE 2.1   Isolation distances required by USDA for producing foundation seed used for seed increase. Note that this is not a complete list of all crops

Crop Species[a]	Distance, ft	Maximal Proportion Contaiminated, %[b]
No isolation required:		
Barley	0	0.05
Bean, field and garden	0	0.05
Broad bean	0	0.05
Cotton	0	0.03
Flax	0	0.05
Millet, selfed	0	0.05
Mung bean	0	0.10
Oat	0	0.02
Pea, field	0	0.50
Peanut	0	0.10
Soybean	0	0.10
Triticale	0	0.05
Wheat	0	0.50
Isolation Required:		
Alfalfa	600	0.10
Buckwheat	660	0.05
Clover, < 2 ha	600	0.10
Clover, > 2 ha	900	0.10
Corn	660	0.10
Crown vetch, < 2 ha	200	0.10
Crown vetch, > 2 ha	900	0.10
Grasses, cross-pollinated	900	0.10
Grasses, selfed	60	0.10
Lespedeza	10	0.10
Millet, cross-pollinated	1,320	0.005
Mustard	1,320	0.05
Okra	1,320	0.0
Onion	5,280	0.0
Pepper	200	0.0
Rape, cross-pollinated	1,320	0.05
Rape, selfed	660	0.05
Rice	10	0.05
Rye	660	0.05
Safflower	1,320	0.01
Sorghum	900	0.005
Sunflower	2,640	0.02
Tobacco	150	0.01
Tomato	200	0.0
Trefoil, birdsfoot	600	0.10
Vetch	10	0.10
Vetch, milk	600	0.05
Watermelon	2,640	0.0

[a]Common name.

[b]Maximal percentage produced from pollen outside plot.

Source: From regulations listed in table 5 of USDA (1994a).

come from outcrossing or cross-fertilization among plants via wind polli-nation (Arriola and Ellstrand 1996; Ellstrand and Foster 1983). Oilseed rape (canola) also produces a mixture of selfed and outcrossed seeds, whereas corn, carrot, sunflower, poplar, radish, strawberry, clover, and many other species produce most of their seeds by outcrossing rather than selfing (for example, Brown et al. 1985; Free 1970; Richards 1986). By definition, outcrossers disperse pollen to other plants and therefore need greater isolation distances than selfing species in situations where pollen dispersal is considered undesirable (table 2.1). Wind-pollinated outcrossers, such as many trees and grasses, tend to produce larger quan-tities of pollen (causing problems for people with pollen allergies) than similar-size plants that are pollinated mainly by insects and other ani-mals. Pollen from animal-pollinated plants is not as buoyant in air cur-rents, and very little is shed into the air; these types of pollen adhere to the pollinator's body. Pollen grains of a few species, such as sugar beet and oilseed rape, are dispersed by both wind and insects (Cresswell et al. 1995; Free et al. 1975; McCartney and Lacey 1991).

The amount of pollen dispersed from outcrossing plants declines rap-idly as a function of distance from the source (section 2.6.2, figure 2.2). In the corn example, the amount of pollen deposited at 60 m was only 0.02% of the amount deposited 1 m from the edge of the crop, but this still represents 2,500 pollen grains per square meter (Raynor et al. 1972). Most pollen is deposited within a few meters of its source, but a small propor-tion can be carried more than 1 km away (for example, Arias and Rieseberg 1994; Kirkpatrick and Wilson 1988; Klinger et al. 1992). Such long-distance gene flow is not easy to measure, because very large samples are needed to detect low-probability events. In addition, the extent of long-distance gene flow is highly variable and depends on local condi-tions, the relative sizes of donor and recipient populations, and synchrony of flowering. Once pollen from a crop has spread to wild plants, further gene flow occurs in a ripple effect through both pollen and seed dispersal.

2.7.2   Crop-to-Wild Gene Flow

The escape of novel resistance traits into free-living populations of wild relatives is often cited as an undesirable consequence of growing transgenic crops on a commercial scale (for example, Bergelson et al. in press; Ellstrand and Hoffman 1990; NRC 1989; Parker and Kareiva 1996; Raybould and Gray 1998; Rissler and Mellon 1996; Snow and Morán-Palma 1997; Tiedje et al. 1989; Van Raamsdonk and Schouten 1997). Con-cerns arise mainly when novel, beneficial traits have the potential to spread to wild populations and cause them to become more invasive and difficult to control. Naturally-occurring damage by insects and diseases can regulate the density and distribution of populations of wild relatives, so release from these pressures by genes from a cultivated relative could potentially make weeds more troublesome in agricultural fields. Moreover, many wild or weedy relatives also occur in other habitats, including rangeland, roadsides, forests, wetlands, and other natural areas where they are not controlled with weed-management techniques, such as the use of herbicides, mowing, and tilling.

Conventional crop genes have spread to wild populations in the past (for example, Arriola and Ellstrand 1996; Ellstrand et al. 1999; Linder et al. 1998; Small 1984), but this process has not been studied by weed scientists, and it is not known whether it has facilitated evolutionary adaptations in weedy relatives (Small 1984; Snow et al. 1998). Circumstantial evidence suggests that several weedy species such as Johnson grass have become more robust and abundant as a result of hybridization with crops (NRC 1989). Known problems have involved hybridization between a weed and a crop, often due to range expansions of one species or the other (Ellstrand et al. 1999; Rissler and Mellon 1996), rather than a transfer of few novel crop genes within an existing crop-weed complex. A well-documented example of a new weed problem occurred in France: a wild gene for early bolting spread from sugar beet seed nurseries. Seeds from those nurseries were then grown by farmers and new weed populations resulted. This led to serious economic losses (Boudry et al. 1993). Scientists have viewed wild relatives mainly as sources of genes that could affect crops, either as useful germplasm or as undesirable pollen contamination; thus, there is little empirical evidence of whether wild species have benefited or not from crop-to-wild gene flow.

Recently, weed scientists have redoubled their efforts to quantify genetic diversity in weed populations and to gain a better understanding of how weeds adapt to changing conditions. The commercialization of transgenic crops has provided added incentives for studies on crop-to-wild gene flow because some transgenic phenotypes have never occurred in wild relatives of certain crops. For example, wild sunflowers lack protection against their seed-eating insects and a fungal disease known as *Sclerotinia* rot (Seiler 1992). If transgenic methods introduced protectants against these pests into cultivated sunflowers, the genes that code for these protectants could move into wild sunflowers, perhaps increasing their ecological fitness and abundance. Likewise, genetic resistance to the herbicide glyphosate (Roundup™) has not been found in the germplasm of crop relatives, but this trait could be transmitted to wild or weedy relatives of transgenic, glyphosate-resistant crops. The use of a wider and more effective variety of transgenic methods could increase the rate at which weedy species acquire novel types of strong, single-gene resistance via crop-wild hybridization.

A first step toward assessing the consequences of crop-to-wild gene flow is to determine which cultivated species are capable of crossing with wild relatives (table 2.2). Corn, soybean, tomato, and cotton do not escape cultivation, nor do they have many close wild relatives within the continental United States (except for cotton in southern Florida). In contrast, some cultivated species survive in feral populations or cross with wild plants of the same species (for example, carrot, sunflower, squash, rice, poplar, certain grasses, oilseed rape, radish, and beet). Many crops have weedy ancestors or occur as weeds in portions of their worldwide distribution (Colwell et al. 1985; Ellstrand et al. 1999; Small 1984). Hybridization between different species or genera is sometimes possible, especially if they share a close, common ancestry. Spontaneous hybridization can occur even when the two taxa have unequal chromosome numbers, which typically creates problems during cell division (for example, Mikkelsen et al. 1996; Zemetra et al. 1998). In fact, new molecular methods have shown that gene flow among related species is much more common than was previously thought (for example, Arnold 1997). Genes often spread among species without being detected by visible indicators, so genetic markers, such as unique DNA fragments are needed to identify current and historical patterns of crop-to-wild gene flow. Recent studies have augmented the number of plant groups in which interspecific hybridization is known to occur naturally, as in oaks, birches, orchids, irises, and wild gourds. The information derived from studies of ongoing gene flow could also be used to understand what types of genes and traits move between these species, thus providing a basis to predict whether specific traits would raise concern if the genes/traits were introduced and transferred.

Early generations of crop-wild hybrids generally have lower fertility than their parents, but there can be high variability among the fertility of the hybrids, with some having high fertility. Even those which produce few viable gametes can transmit crop genes to later generations (a process known as introgression). With each successive generation of backcrossing with wild genotypes, the crop's contribution to the plant's total genome is reduced by 50% (figure 2.3) and the progeny become more similar to wild genotypes. After two or three generations of backcrossing, plants with a crop ancestor can be just as competitive and successful as wild plants (for example, Snow et al. 1998). Moreover, the fitness of *first*-generation crop-wild hybrids is sometime as great as the fitness of wild plants (for example, Arriola and Ellstrand 1997; Klinger and Ellstrand 1994).

The frequency of a given crop gene in a wild population depends on many factors, including the rate at which it is introduced into the population; temporary fitness barriers, if any, in the first and early backcross generations; possible fitness costs associated with the gene itself; and

TABLE 2.2   Examples of Commercially Important Species That Can
Hybridize with Wild Relatives in the Continental United States

Family and Cultivated Species[a]	Wild Relative[a]
Apiaceae	
Apium graveolens (celery)	Same species
Daucus carota (carrot)	**Same species (wild carrot)**
Chenopodiaceae	
Beta vulgaris (beet)	*B. vulgaris* var. *maritima* (<u>hybrid</u> is a weed)
Chenopodium quinoa (quinua, a grain)	*C. berlandieri*
Compositae	
Chicorium intybus (chicory)	<u>**Same species**</u>
Helianthus annuus (sunflower)	<u>Same species</u>
Lactuca sativa (lettuce)	*L. serriola* (wild lettuce)
Cruciferae	
Brassica napus (oilseed rape;canola)[b]	<u>Same species</u>, *B. campestris*, *B. juncea*
Brassica rapa (turnip)	<u>Same species</u> (= *B. campestris*)
Raphanus sativus (radish)	<u>Same species</u>, *R. raphanistrum*
Cucurbitaceae	
Cucurbita pepo (squash)	<u>Same species</u> (= *C. texana*, Wild squash)
Ericaceae	
Vaccinium macrocarpon (cranberry)	Same species
Vaccinium angustifolium (blueberry)	Same species
Fabaceae	
Trifolium spp. (clover)	<u>Same species</u>
Medicago sativa (alfalfa)	Same species
Hamamelidaceae	
Liquidambar styraciflua (sweetgum)	Same species
Juglandaceae	
Juglans regia (walnut)	*J. hindsii*
Liliaceae	
Asparagus officinalis (asparagus)	Same species
Pinaceae	
Picea glauca (spruce)	Same species
Poaceae	
Avena sativa (oat)	***A. fatua*** (wild oats)
Cynodon dactylon (bermuda grass)	**Same species**
Oryza sativa (rice)	**Same species** & others (red rice)
Saccharum officinarum (sugar cane)[c]	***S. spontaneum*** (wild sugarcane)
Sorghum bicolor (sorghum)	***S. halepense*** (johnsongrass)
	<u>Same Species</u>[d] (shattercane)
Triticum aestivum (wheat)	<u>*Aegilops cylindrica*</u> (jointed goatgrass)[e]

TABLE 2.2 *Continued*

Family and Cultivated Species[a]	Wild Relative[a]
Rosaceae	
Amelanchier laevis (serviceberry)	Same species
Fragaria sp. (strawberry)	*Fragaria virginiana*
Rubus spp. (raspberry, blackberry)	<u>Same species</u>
Salicaceae	
Populus alba x *P. grandidentata* (poplar)	<u>*Populus*</u> spp.
Solanaceae	
Nicotiana tabacum (tobacco)	Same species
Vitaceae	
Vitis vinifera (grape)	<u>*Vitis*</u> spp. (wild grape)

[a]Wild relatives recognized as weeds (unwanted species in agricultural or other habitats) are underlined; those also included in the worst 100 weeds worldwide (Holm et al. 1997) or Federal Noxious Weed List are in boldface. This list is not exhaustive, especially for landscaping and forage species. For some cultivars the extent of hybridization has not been studied.

[b]Also hybridizes with *Raphanus raphanistrum*, but evidence to date suggests that crop chromosomes do not recombine with wild genome and are lost after several generations (Chevre et al. 1997; Chevre et al. 1999).

[c]Cultivated sugar cane does not need to flower before harvest, but hybrids can occur (Stevenson 1965).

[d]From Burnside (1968).

[e]From Zemetra et al. (1998).

Source: Adapted from Snow and Morán-Palma (1997).

possible benefits of the gene for the plant's survival and reproduction. Fitness costs associated with transgenes may be small or absent when a seed is ready to be sold, because seed companies choose breeding lines with the best transformation events available (for example, Fredshavn et al. 1995). When several transgenes are inserted together as tightly linked traits (inherited as a unit), the *combined* ecological costs and benefits of these traits will determine whether and to what extent a wild relative's fitness is enhanced. If one transgene confers resistance to a common herbicide, such as glyphosate, selective pressures favoring plants with the transgene will be very strong when that particular herbicide is used, thereby increasing frequencies of the linked transgene in weed populations. Eventually, wild species will be able to acquire additional beneficial transgenes that are released in different cultivars, and the new traits could accumulate in wild populations.

The ecological and evolutionary benefits conferred by crop genes that

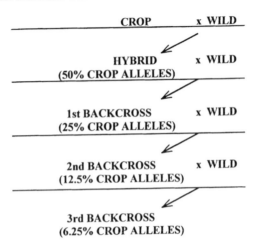

FIGURE 2.3   Dilution of crop alleles in plant genotypes after spontaneous hybridization and backcrossing with wild plants. Two parent plants are shown on each line (for example, crop x wild), with an arrow pointing to their offspring.

enter wild populations are difficult to evaluate. This is an issue that lies at the crux of concerns about gene flow from transgenic and conventional crops. We know little about the extent to which insects and diseases limit wild, weedy populations that are sexually compatible with cultivated species. Critics of biotechnology argue that the spread of beneficial traits could quickly lead to the spread of weeds; advocates of transgenic crops maintain that this risk is small or nonexistent. Empirical data with which to address the question are lacking. Many publications describe proposed methods for evaluating the effects of beneficial crop genes on the dynamics of wild, weedy populations (for example, Crawley et al. 1993; Kareiva et al. 1996; Rissler and Mellon 1996). However, inadequate funding and restrictions on trial releases of transgenic pest-protected plants have hampered opportunities to carry out this important research before commercialization (Wrubel et al. 1992; Purrington and Bergelson 1995). A few studies have focused on how novel genes affect plant fitness, (that is, the relative survival and reproduction of new genotypes) (for example, Bartsch et al. 1996; Stewart et al. 1997), but it is much more difficult to determine how new traits will affect the geographic distribution and local abundances of a given species. The latter requires a combination of field studies, field experiments, and mathematical modeling (for example, Rees and Paynter 1997).

Until better data are available, it will be necessary to rely on general ecological and agricultural knowledge to predict the consequences of commercial-scale, crop-to-wild gene flow from pest-protected plants. The common perception that most wild plants are adapted to tolerate or resist pests is misleading in that even species that appear to be protected by mechanical or chemical defenses can be decimated by specialist herbivores or new diseases. Many studies have demonstrated that herbivores, especially insects that specialize on flowers and seeds, can regulate wild plant populations (for example, Guretzky and Louda 1997; Rees and Paytner 1997; Waloff and Richards 1977). In contrast, much less is known about the extent to which diseases affect the population dynamics of wild plants (Burdon 1987; Burdon and Jarosz 1988). The impact of diseases is much more obvious in cultivated species, especially genetically uniform plants grown at high densities. However in natural populations, where nonspecific plants are often scattered and genetically diverse, disease could still be important. Wild plants are susceptible to many native, introduced, and newly evolved diseases, but we know little about how often they are exposed to these pathogens and whether such exposure constrains population growth. Although general surveys of disease frequencies in wild populations are helpful, they might miss infections that occur inconspicuously (in seeds, seedlings, or roots) or infrequently, and those which are so virulent that they cause local extinctions.

Because of the uncertainties described above, it is premature to predict the ecological impacts of gene flow from transgenic pest-protected plants. Meanwhile, regulatory decisions must be made in a timely fashion. It seems unlikely that the transfer of one or two novel crop genes for pest-protection would transform a wild species into a problematic weed, although in some cases unwanted population increases of weedy species could result. Moreover, the cumulative effects of beneficial crop genes could potentially lead to expensive and ecologically damaging problems in weeds that are already difficult to control, such as Johnson grass (*Sorghum halepense*). In the future, additional phenotypic traits might include broad-spectrum resistance to insects or diseases and greater tolerance of cold, drought, salinity, nutrient scarcity, or acidic soils. Such traits could be more advantageous to wild relatives than those now in use.

Consequences of gene flow other than weediness are also perceived to be detrimental to preserving biodiversity. For example, the spread of transgenes to wild relatives that are rare or endangered is sometimes considered as a potential ecological risk, especially in regions that are centers of diversity for crop relatives (for example, Rissler and Mellon 1996). However, if a portion of wild species's genome is being exchanged for new genetic material by hybridization, this process will occur whether or not the crop is transgenic. The transfer of resistance traits to rare wild relatives is unlikely to exacerbate population declines or lower genetic diversity in these species, given that transfer of traits between crops and wild relatives is an ongoing process. However, researchers and regulators should be aware of these and other potential unintended consequences of gene flow from transgenic pest-protected plants (for example, the hypothetical spread of a Bt gene from an ornamental species to the only host plant of a rare beetle).

Eventually, it could be possible to reduce gene flow from cultivated plants with various containment methods. If a transgene is inserted into the chloroplast genome, which is usually not transmitted by pollen, spread of the transgene into natural populations could be delayed (Daniell et al. 1998) but would not be avoided entirely (Stewart and Prakesh 1998). Gressel (1999) suggested linking transgenes to genes for traits that are detrimental to weeds but not crops (for example, a gene for lack of seed dormancy). Again, this would slow down the process of introgression but would not necessarily provide complete containment of ecologically beneficial transgenes. A more effective method of confining gene flow may be to use a system in which transgenic seeds are killed just before they mature, so that they can still be harvested but will not germinate (US Patent No. 5,723,765, "Control of plant gene expression"). Popularly known as "terminator" technology, this potential method of containment is highly controversial and has not yet been used commercially. In the future, biotechnology companies might develop transgenic plants with inducible pest resistance traits that require a chemical spray to be "turned on." If these systems work as predicted, resistance genes that spread to wild species would not be expressed in the wild species unless the spray was used, and therefore, would be extremely unlikely to contribute to the wild species' invasiveness. Because of the concerns raised about using chemicals to turn on genes in the environment, it is unclear whether or not such technology will be acceptable. However, if this strategy is pursued on a commercial scale, it could greatly reduce the need to regulate transgenic crops based on weedy-relative considerations.

2.7.3   Crop-to-Crop Gene Flow

Crop breeders try to minimize gene flow into their breeding lines by using crop-specific isolation distances (for example, Bateman 1947a, b, c). Seeds used for commercial sale are often mandated to have less than 0.1% contamination from phenotypically different sources (table 2.1). Less stringent genetic isolation is required for producing certified seeds to sell to farmers, and only rarely is any isolation required when farmers grow a crop for harvest. Now, however, organic farmers are concerned about gene flow from transgenic plants as organic standards may forbid the use of transgenic products. Under proposed standards, organic farmers are not allowed to sell transgenic food, and a tolerance for very low levels of contamination by transgenic DNA has not been established (NOSB 1996). The committee knows of no scientific evidence that crop-to-crop gene flow has caused health or environmental risks to date, but we recognize that consumer choices are involved. Nonorganic farmers might also face problems with crop-to-crop gene flow and undesirable seed characteristics—for example, if transgenic crops are commercially developed for pharmaceutical or industrial compounds, or when farmers' fields are inspected in search of the unauthorized use of patented transgenic genotypes. Clearly, these hypothetical concerns apply to a variety of transgenic crops. Contamination with pollen from other farms is likely to be very low in most cases, as illustrated in figure 2.2 and table 2.1 and described further below.

Crop-to-crop gene flow is unlikely to occur over long distances in species with high selfing rates, although even rice, sorghum, and wheat are known to hybridize occasionally with nearby plants (Arriola and Ellstrand 1996; Langevin et al. 1990; Seefeldt et al. 1998; Zemetra et al. 1998). Pollen from outcrossing species could easily move 600 ft or more among plantings (table 2.1). In a study of oilseed rape fields in Scotland, researchers documented a small amount of pollen dispersal as far as 3 km from the crop, indicating that farm-to-farm spread of transgenes was common (Thompson et al. 1999). In that region, over 60% of the crop plantings occurred within 100 m of another crop planting. Gene flow into row crops can also occur via volunteer crop plants that survive to reproduce and via feral plants that start new populations, for example, when sunflower or oilseed rape seeds are inadvertently spilled along roadsides. Even for selfing species, total containment of crop genes is not considered to be feasible when seeds are distributed and grown on a commercial scale. The inevitability of gene escape has been recognized by federal agencies, and regulatory decisions are based on it.

In some situations, unwanted levels of contamination can be avoided by altering the distances among plots, using border rows to intercept unwanted pollen, or planting large acreages to minimize edge effects (for example, figure 2.2 and table 2.1; Hokanson et al. 1997; Morris et al. 1994). These approaches have been discussed and recommended by the USDA for field testing of transgenic plants products prior to the regulatory approval of these products (USDA 1993; USDA 1996a). Figure 2.4 illustrates the general principle that small populations are more susceptible to contamination from long-distance pollen sources than are large populations (for example, Ellstrand and Elam 1993; but see Klinger et al. 1992). For outcrossing cultivated plants, contamination of less than 1% of the seeds in a given crop can often be prevented by isolation distances of over 1 km, but the threshold distance depends on the species, the relative sizes of donor and recipient plantings, use of border rows, and overlap in flowering times.

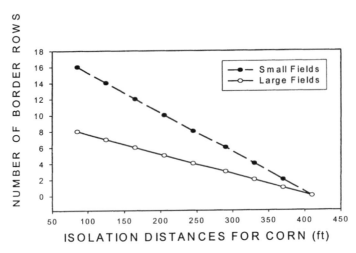

FIGURE 2.4.   Number of border rows recommended to keep "off-type" seeds in corn fields under 0.5% as a function of isolation distance from other sources in corn fields. These requirements apply to contaminating pollen from sources with kernels that closely resemble the crop; therefore, "under 0.5%" does not refer to actual gene flow, but rather to acceptable levels of gene flow for producing certified seeds of standard quality. Small fields are defined as less than 20 acres; large fields are more than 20 acres. Source: From seed certification regulations listed in USDA 1994a.

2.7.4   Summary

In light of the above discussions, the committee found that

Pollen dispersal can lead to gene flow among cultivars and from cultivars to wild relatives, but only trace amounts of pollen are typically dispersed further than a few hundred feet.

The transfer of resistance traits to weedy relatives could potentially exacerbate weed problems, but such problems have not been observed or adequately studied.

The committee recommends that

Criteria for evaluating the merit of commercializing a new transgenic pest-protected plant should include whether gene flow to feral plants or wild relatives is likely to have a significant impact on these populations.

In order to study the effects of gene flow, the committee recommends areas for research in chapter 3 and the executive summary (sections 3.4 and ES.5.2).

2.8 AGRONOMIC RISKS ASSOCIATED WITH VIRUS-RESISTANT CROPS

Agronomic risks are defined as those related to quality or productivity of a modified crop. A number of issues and concerns emerge when agronomic consequences of using transgenic pest-protection strategies against viruses are considered. These center on emergence of new or novel viral strains, introduction of new transmission characteristics, and changes in susceptibility to heterologous viruses. Some of the concerns, such as the question of new virus emergence, also have relevance to ecological risks.

2.8.1 Evolution of Resistance to Pest-Protected Crops

The emergence of strains of pathogens that overcome plant-genetic resistance or other disease-control methods has been and probably always will be a problem in agriculture. Indeed, this problem is common to all plant and animal hosts for which pathogens exist. Traditionally the problem has been managed by development of multiple strategies for disease control (genetic and other control measures), surveillance of pathogen activity and strain development, deployment of new pest-protected germplasm in response to emerging pest strains, and development and use of longer-lasting forms of pest-protection. With transgenic pest-protected plants that express a pest-protection gene transferred from another plant, the selective pressure for development of resistance-breaking strains should be qualitatively similar to the selective pressure associated with conventionally bred pest-protected plants. In section 2.9, these issues are discussed.

2.8.2 Risks Posed by Virus-Derived Transgenes

Recombination Between Transgenes and Viral Pathogens

Recombination between a virus-derived transgene and a virus during plant infection has been suggested as a potential source of novel virus strains with enhanced virulence characteristics. From an evolutionary perspective, new viruses emerge through gradual accumulation of point mutations and major acquisition or deletion of genetic material. New genetic material can be incorporated by recombination with nucleic acids from the host or from other viruses during coinfections (Koonin and Dolja 1993). Several laboratory studies have shown that viruses can recombine with homologous transgene sequences (Borja et al. 1999; Greene and Allison 1994; Wintermantel and Schoelz 1996). Those studies required moderate or heavy selection pressure to detect the recombinant viruses. In all cases examined, only homologous sequences were exchanged. No experimental data indicate that recombination can occur between virus genomes and transgene sequences that are derived from distantly related or unrelated viruses.

Two points should be considered in assessing risks that may be posed by virus-transgene recombination. First, will large-scale plantings of transgenic material increase the risk of recombination above the preexisting risk due to the widespread occurrence of mixed infections? Mixed infections by related and unrelated viruses are common in natural and agricultural ecosystems. For example, mixed infections by two or more viruses were detected in 64% and 90% of peppers surveyed in three California counties in 1984 and 1985, respectively (Abdalla et al. 1991). Mixed infections provide a continuous opportunity for intervirus recombination. Either because intervirus recombination occurs with such low frequency or because new recombinants so rarely have a competitive advantage, the new viruses have not been detected in agricultural settings. One could argue that past and current agricultural practices have provided a fertile environment to spawn novel recombinant viruses with virulent properties, but these viruses have not been observed.

Second, can transgenes be engineered to reduce or eliminate the risk that recombination will spawn new pathogens? Evidence suggests that elimination of genome replication-control sequences from transgenes can limit recombination and therefore risk (Greene and Allison 1996). Furthermore, strategies to produce resistance-mediating transgenes that encode nonfunction proteins or no protein can be used effectively against viruses. For example, resistant plants that express nontranslatable RNA can confer immunity through induction of post-transcriptional gene silencing (Kasschau and Carrington 1998; Lindbo and Dougherty 1992).

Transcapsidation and Gain-of-Transmission Characters

In the process of encapsidation, a virus genome is packaged in a shell of self-encoded coat proteins after it is replicated. Although encapsidation of one virus genome by the coat protein from a different virus (transcapsidation) is well documented (Matthews 1991), it is highly unlikely that functional coat proteins expressed in transgenic plants pose a significant risk of expansion of host range to new crop or non-crop hosts. Transcapsidation does not involve exchange of genetic material, meaning that any unique insect vectoring properties of a transcapsidated virus genome will not be inherited.

Synergism Between Viral Transgenes and Heterologous Viruses

Mixed infections by viruses can sometimes lead to a synergistic disease syndrome that is more severe than that caused by either virus individually (Matthews 1991). In cases of synergism involving the potyvirus family of viruses, the region of the viral genome that causes exacerbation of disease codes for a protein termed HC-Pro (Pruss et al. 1997). The synergism effect appears due to the natural role of HC-Pro as a suppressor of the gene-silencing response (Scheid 1999). Thus, plants are unable to mount an effective defense response to infection. Indeed, transgenic plants that express potyvirus genome segments that include HC-Pro exhibit more severe symptoms when inoculated with heterologous viruses (Pruss et al. 1997). No data indicate that expression of viral coat protein or replicase proteins enhances the virulence of heterologous viruses.

The problem of synergism is manageable through avoiding the use of functional transgenes that encode defense-suppressor substances or pathogenicity-enhancer substances. In addition, the normal process of testing in breeding programs that seek to incorporate natural or transgenic resistance traits will reveal the extent to which the virulence of heterologous viruses is exacerbated. Thus, it is highly unlikely that transgenic plants with general hyper susceptibility characteristics will pass through a breeding program to commercialization.

2.8.3   Summary

In light of the above analysis, the committee found that

Most virus-derived resistance genes are unlikely to present unusual or unmanageable problems that differ from those associated with traditional breeding for virus resistance.

Case studies of virus resistant squash and papaya are presented in chapter 3 (section 3.1.4).

2.9   PEST RESISTANCE TO PEST-PROTECTED PLANTS AND RESISTANCE MANAGEMENT

In this section, the ability of pests to adapt and develop resistance to transgenic or conventional pest-protected plants will be discussed, and resistance management strategies to abate this development and their scientific basis will be presented.

2.9.1   Pest Resistance to Control Tactics

The history of agricultural pest management has demonstrated that insects, weeds, and microbial pathogens have the evolutionary potential to overcome or circumvent most control tactics imposed by farmers (Barrett 1983; Green et al. 1990; Gould et al. 1991). That over 400 insect species have become resistant to at least one insecticide (Georghiou 1986) is often cited as evidence of the genetic ability of arthropods to evolve resistant strains. In addition, weeds and pathogens also have an impressive record of successful adaptation to control measures (Green et al. 1990). Although the number of cases of resistance by weeds to herbicides is smaller than that of insects, the percentage of weed species that has developed resistance is greater than that of insects (Gould 1995). It is well documented that microbial pathogens can successfully adapt to crop cultivars that are bred to resist specific diseases (Lamberti et al. 1981). In examining a random selection of 63 cases of viruses that live in association with specific plant hosts, Fraser (1990) found that in 28 of the cases there were good data indicating that adapted isolates of the virus existed (in only five cases was there good evidence that there had been no adaptation by the virus). Fungal and bacterial adaptation to pest-protected cultivars has caused serious crop losses. In many cases, pathogen resistance has occurred less than 5 years after a classically bred resistant cultivar was released for commercial use (see section 3.1.1). Experience with both insect and pathogen adaptation to genetically modified pest-protected (GMPP)[6] plants indicates that the more intensively a control tactic is used, the more rapidly pests will adapt to it (Gould et al. 1991). History also indicates that pests adapt more rapidly to some types of GMPP plants than to others (Lamberti et al. 1981).

If a GMPP cultivar is lost because the target pest adapts to the cultivar, replacing the cultivar with a new GMPP cultivar can have a number of associated costs. Even if new GMPP genes are available, moving those genes into a modern cultivar is expensive. Although the health and environmental safety of the plant protectant in the new cultivar can be tested in laboratory experiments, the new cultivar will need to be monitored for impacts that could not have been detected in the laboratory experiments. If new pest-protection genes are not available, farmers might need to move back to reliance on broad-spectrum pesticides. Decreasing the rate at which target pests adapt to GMPP cultivars can therefore produce societal benefits.

---

[6]As a reminder to the reader, GMPP plants include both transgenic and conventional pest-protected plants. See section ES.3.

2.9.2   Decrease in the Rate of Resistance Evolution

Concern over the risk of pest resistance to conventional pesticides led to development of a relatively new field of applied science, called pest-resistance management (NRC 1986). The goal of this field is to determine approaches for developing and deploying pest control tactics in ways that maximize long-term benefits. Pest-resistance management is grounded in concepts and empirical findings from the basic sciences of quantitative genetics and population genetics (NRC 1986). In this regard, it is very similar to the applied science of classical crop breeding. These fields of inquiry rely heavily on statistical inference. A theoretical population geneticist or a crop breeder is therefore unlikely to make a deterministic prediction about the outcome of a natural evolutionary event or the exact characteristics of his or her next new cultivar. For the same reason, scientists investigating pest-resistance management tactics are reluctant to provide regulatory officials or farmers with exact predictions about how many years it will take for a specific pest to adapt to overcome a proposed resistance management plan. However, they can provide information on which of a number of approaches to development and deployment of transgenic and conventional pest-protected plants is likely to be most successful in decreasing the rate of pest evolution to adapt to those plants.

Quantitative comparisons of resistance management approaches for crops protected against insect damage began in 1986 (Cox and Hatchett 1986; Gould 1986a, b). A list of potential approaches has since been developed (Gould 1988a; McGaughey and Whalon 1992; Roush 1997; Tabashnik 1994). Some of the general approaches for resistance management for insect pests are as follows:

- High dose of a single contained toxin in most plants, with some plants producing no toxin at all and thus serving as a refuge (approach 1).
- Multiple toxins at high (or in some cases moderate) doses in most plants, with some nontoxic plants serving as a refuge (approach 2).
- GMPP plants with low doses of a toxin that only slow the growth of the pest, so that pest population growth decreases and natural enemies can become more effective (approach 3).
- Development of GMPP plants that produce the toxin only when and where it is most critical to protecting the plant (approach 4).

Much of the research aimed at developing these approaches and assessing their expected impacts has focused on transgenic pest-protected plants that produce Bt toxins.

A high dose of a toxin has been defined by the EPA Science Advisory

Panel (SAP 1998) as an amount that is 25 times the amount needed to kill 99% of susceptible insects. Empirical information on development of insect genetic adaptation to Bt toxins indicates that such a high dose will kill most partially adapted insects in a pest population (such as heterozygotes). The result is interruption of the typical stepwise process of evolving from susceptible populations, to one that is partially adapted, to fully adapted to the GMPP plant. An analogy can be made to the use of antibiotics to treat human pathogens. The utility of decreasing the survival of partially adapted human pathogens has long been recognized by medical researchers and physicians who routinely recommend that their patients continue to take antibiotics past the period when most of the infectious organisms have been killed. The prolonged treatment period ensures that partially adapted target pathogens will also be killed and so not be transmitted to other people.

Many researchers have examined field and laboratory insect populations in an attempt to understand the mechanisms of insect adaptation to Bt toxins (Tabashnik 1994; Gould and Tabashnik 1998). Results indicate that either multiple genes or single recessive genes are needed to confer full adaptation to a high dose of Bt although partial resistance can be confered by a dominant gene. A very low proportion (1 out of a million, to 1 out of 1000) of fully adapted insects is expected to exist in a population before the population is exposed to Bt. If all host plants on a farm produced the high dose of Bt, only those few insects with the right gene combinations would survive. If they then mated with each other, their offspring would be fully adapted, and the pest population would no longer be affected by the GMPP crop. The planting of nontoxic host plants (refuges) is designed to make sure that a relatively large number of totally susceptible insects are produced on each farm, compared with the few fully adapted insects produced. As long as this refuge is maintained, almost all fully adapted insects produced in the Bt crop are expected to mate with susceptible insects. The offspring of these matings will not have the proper combination of genes needed to be fully adapted, so the evolutionary process is again interrupted. Many researchers expect use of the refuge in combination with plants that produce a high dose of the toxin to increase the time needed for insect adaptation by a factor of 10 (for example, from eight years to more than 80 years) if properly implemented (Gould 1998; Roush 1997; Tabashnik 1994).

By using transgenic or conventional pest-protected plants that contain high levels of multiple toxins with high doses (approach 2), the chance that insects will have the proper gene combination to be fully adapted is further decreased compared to the case where only one toxin is produced by the plant. It also increases the efficiency with which refuge-produced insects can break up combinations of resistance genes from the few pests that happen to carry the proper gene combination (Gould 1986a, b; Roush 1997).   The development of resistance may be delayed by the use of several toxins with different modes of action (Zhu et al. 1994; Jach et al. 1995). The toxins could arise from a combination of conventional breed-ing and transgenic techniques. However, even if the target sites of two toxins differ, there is still the possibility of cross resistance if the two toxins can be detoxified by the same enzymes. The new high-dose refuge approach (approach 1) has been the most widely accepted tactic for resis-tance management of target pests of transgenic or conventional pest-pro-tected plants. Approaches 3 and 4 also have potential applicability. Ap-proach 3, which relies on an interaction between the GMPP plant and natural enemies, is expected to decrease the rate of evolution of adapta-tion because it does not result in a major decrease in the fitness of either susceptible or adapted pests. Companies have not embraced this ap-proach, because it cannot always be relied on to protect the crop, and they may have liability for control failures. There have also been concerns that the approach might not always inhibit evolution of adaptation to the pest-protected plant (for example, Johnson et al. 1997a, b). Approach 4 would also decrease the rate of evolution of resistance because only the fraction of the pest population that feeds on the protected-plant parts would be killed. This general approach could be useful if correctly implemented, but technological and ecological problems must be solved before it can be used (Roush 1997).

The high-dose approach is feasible with Bt toxins because even at high doses no health or environmental problems have been reported in commercially grown varieties. Also, crop yield has not been reduced by production of high doses. That might not be the case with some other plant-protection mechanisms.

Much of the theory of resistance management for GMPP plants has been developed for diploid, sexually reproducing organisms (NRC 1986; Roush and Tabashnik 1990). The theory is therefore only partially appli-cable to viruses, bacteria, and even a large group of insects that have different means of reproduction.

Plant pathologists have long been concerned with viral, fungal, and bacterial adaptation to conventional pest-protected plants. In the 1950s they developed the concept of GMPP plants having either vertical or horizontal resistance to pathogens (Van der Plank 1963). Vertical resis-tance typically involved single plant genes that were initially very effec-tive at mitigating a disease but were expected to be evolutionarily over-come by rapid genetic shifts in the pathogen (Lamberti et al. 1981). Horizontal resistance was typically controlled by many genes, offered lower but adequate suppression of the target pathogen, and was expected to be more durable (recalcitrant to pathogen adaptation). This system for stereotyping GMPP plants had some predictive power, but later investigators found that there were too many exceptions (Lamberti et al. 1981). Durability of a specific GMPP plant is now typically judged in retrospect on the basis of its long term performance (Johnson 1981).

There have been some recent attempts to use population-genetics theory for developing and deploying conventional pest-protected plants in ways that slow pathogen adaptation (Burdon et al. 1994; Lannou and Mundt 1996 and 1997; Mundt 1990; Zeigler 1998), but it has not become common practice. Instead, many current pathology programs for production of GMPP plants emphasize continual discovery of new resistance genes so that breeding programs can stay a step ahead of an evolving pathogen (McIntosh and Brown 1997).

Researchers developing engineered pathogen-resistant plants have also been concerned with pest adaptation (Beachy 1997). Although new molecular approaches could lead to plants that offer a greater evolutionary challenge to pathogens (Beachy 1997; Bendahmane et al. 1997), little empirical or theoretical work has been aimed at determining how to produce durable engineered pathogen resistance (but see Qiu and Moyer 1999).

2.9.3   Future of Resistance Management for GMPP Plants

EPA has been active in developing resistance management plans for Bt crops. It has developed an internal group of staff to work on the issue and has consulted formally and informally with researchers (Matten et al. 1996; Matten 1998). Researchers and EPA regulatory officials will probably learn a lot of general principles about how to develop and implement resistance management of transgenic pest-protected plants from the continuing work on Bt crops. Much has already been learned from the Bt system regarding theoretical and practical aspects of developing and implementing a resistance management program, but the EPA policy is still evolving (Matten 1998). Each year, new empirical results should provide information on better ways to optimize resistance management for these crops. Therefore, plans implemented today will need to be periodically reviewed for their continued usefulness.

Although EPA has instituted programs and regulations that demonstrate serious concern about insect adaptation to Bt crops (Matten 1998; SAP 1998), it has not indicated concern about virus adaptation to transgenic pest-protected plants with plant-produced viral coat protein. In general, EPA has not commented in formal documents about when it considers pest adaptation to pest-protected plants to be an important public or social problem and when it considers resistance to be only a business problem or an insignificant public or social problem.

There has been considerable public debate about this issue. One opinion is that there is no more reason to institute resistance management for transgenically produced Bt than there is to institute resistance management for conventional pesticides. Others argue that the use of Bt toxins in transgenic pest-protected crops is fundamentally different from the use of chemical pesticides, for a combination of the following reasons:

- Insecticides are typically used only when pest populations increase to the point where substantial yield-losses could occur, so refuges are already present. With transgenic Bt crops, the toxin is selecting for resistance all season long, even during weeks when the pest cannot feed on plant parts that affect crop yield, or during years when pest number are too low to cause yield loss.
- Bt toxins are seen as benign to the environment and public health, and no equally benign replacement product is available.
- Bt toxins are the active component in Bt spray formulations that have been used sustainably by organic and conventional farmers for many years, and this tool could be lost if transgenic Bt crops are not managed correctly.

Many transgenic pest-protected plants of the future may be protected by novel mechanisms and therefore not compromise the utility of plant protectants that are already being used by farmers. In such cases, the company that produces the plant protectant can be seen as the major party affected by pest evolution of adaptation to the company's product. However, there could be cases in which a new transgenic pest-protected plant cultivar is produced by transferring a plant protectant that is already in use to another crop species. The new use could increase the risk that pests will involve adaptation to the plant protectant in all uses. An example might be moving a pathogen-resistance gene from tomato into cotton. If the same pathogen is now controlled by this resistance-mechanism in both crops, the intensity of selection for adaptation could be substantially increased. A resistance management program could be developed in such a situation to ensure that adaptation does not evolve at a rate or in a manner that causes environmental, economic, or health problems.

2.9.4   Summary

In light of the above discussion, the committee found that

Evolution of pest strains that can overcome the pest-protection mechanisms of plants can have a number of potential environmental and health impacts.

For example, adaptation of a pest to an environmentally pesticidal substance produced by a transgenic plant may cause farmers to return to or begin the use of a conventional chemical pesticide with toxic effects on nontarget organisms. Also, adaptation of a pest to one type of transgenic pest-protected plant could result in its replacement with a novel transgenic pest-protected plant for which there is less information regarding health and environmental impacts.

Our understanding of the evolution of adaptation to pest-protected plants is still limited, but there is reasonable expectation that specific approaches to the development and deployment of transgenic pest-protected crops can substantially delay the evolution of pest adaptation. The committee found that

Although EPA has worked actively to develop useful resistance management plans for crops containing *Bacillus thuringiensis* (Bt) toxins, the agency has not articulated a general policy indicating when it believes it should require the development of resistance management plans for specific transgenic pest-protected crops.

The committee recommends that EPA continue to deal seriously with Bt resistance management (section 1.6.1), and it should also begin to consider resistance management strategies for other transgenic pest-protected plants. Specifically,

If a pest protectant or its functional equivalent is providing effective pest control, and if growing a new transgenic pest-protected plant variety threatens the utility of the existing uses of the pest-protectant or its functional equivalent, implementation of resistance management practices for all uses should be encouraged (for example, Bt proteins used both in microbial sprays and in transgenic pest-protected plants).

2.10   RECOMMENDATIONS

- **When the active ingredient of a transgenic pest-protected plant is a protein and when health effects data are required, both short-term oral toxicity and potential for allergenicity should be tested. Additional categories of health effects testing (such as carcinogenicity) should not be required unless justified.**

- **The EPA should provied clear, scientifically justifiable criteria for establishing biochemical and functional equivalency when registrants request permission to test non plant-expressed proteins in lieu of plant-expressed proteins.**

- Priority should be given to the development of improved methods for identifying potential allergens in pest-protected plants, specifically, the development of tests with human immune-system endpoints and of more reliable animal models.

- Criteria for evaluating the merit of commercializing a new transgenic pest-protected plant should include the anticipated impacts on nontarget organisms compared with those of currently used pest control techniques[7] and whether gene flow to feral plants or wild relatives is likely to have a significant impact on these populations.

- If a pest protectant or its functional equivalent is providing effective pest control, and if growing a new transgenic pest-protected plant variety threatens the utility of existing uses of the pest protectant or its functional equivalent, implementation of resistance management practices for all uses should be encouraged (for example, Bt proteins used both in microbial sprays and in transgenic pest-protected plants).

---

[7]Includes both chemical and non-chemical methods which are currently used.

3

Crossroads of Science and Oversight

This chapter focuses on the scientific basis of the oversight of transgenic pest-protected plants. The committee recognizes that there is an urgency to solidify the regulatory framework for transgenic pest-protected plant products because of the potential diversity of novel traits that could be introduced by transgenic methods and because of the rapid rate of adoption of and public controversy regarding transgenic products.

For comparison with transgenic pest-protected plants, a case study concerning conventional pest-protected plants and a discussion of scientific issues surrounding them are presented. Then case studies of transgenic pest-protected plants are discussed; these case studies provide examples of scientific review of transgenic pest-protected plants by federal agencies.

After the case studies, EPA's proposed rule for the regulation of genetically modified pest-protected (GMPP) plant gene products as plant-pesticides is then evaluated in light of the discussion and other scientific criteria. Finally, the committee suggests guiding scientific principles for oversight of transgenic pest-protected plants and sets forth specific research needs.

3.1 CASE STUDIES OF PEST-PROTECTED CROPS AND THEIR OVERSIGHT

3.1.1 Conventional Breeding for Rust Resistance in Wheat

Three main rust diseases affect common wheat (*Triticum aestivum* L.) and durum wheat (*T. durum* Desf.): stem rust (or black rust), caused by

Puccinia graminis Pers. f. sp. *tritici* Eriks and Henn.; leaf rust (or brown rust), caused by *P. recondita* Rob. Ex Desm. f. sp. *Tritici;* and stripe rust (or yellow rust), caused by *P. striiformis* West. All three rust diseases are fungi which are obligate parasites in nature (that is, they require a living host to survive). They all need free moisture for infection, but they have different optimal environmental conditions for disease development, so they often do not damage wheat production concurrently in the same region (Knott 1989). For example, stem rust tends to require higher temperatures than leaf rust, which requires higher temperatures than stripe rust; hence, stem rust is usually more damaging in the northern Great Plains, leaf rust in the southern Great Plains and the East, and stripe rust in the West. Of the three diseases, stem rust can cause the more devastating epidemics; for example, in 1916, a stem rust epidemic was estimated to have reduced total US wheat production by 38% (Loegering 1967). But leaf rust is more common (Schafer 1987; table 3.1). Because wheat is used primarily as a food grain, losses in total production underestimate the true economic loss when wheat is damaged so severely that it must be sold as a feed grain.

Research to Reduce Losses Caused by Wheat Rusts

Losses due to rusts can be reduced by cultural practices, removal of an alternative host, chemical control, and genetic protection from the pathogens (also called host-plant resistance) (Knott 1989; Schafer 1987). The goal is to break the life cycle of the pathogen.

After the 1916 stem rust epidemic, a major barberry-eradication program was started in North America (Roelfs 1982). Roelfs suggested four benefits from the success of the program: disease onset was delayed by 10 days, initial inoculum was reduced, the number of pathogen races was reduced, and the pathogenic races of stem rust were stabilized. The reduction in the number of races and their stabilization were due to eliminating the sexual cycle of *P. graminis*. *P. recondita* also has alternative hosts, but none is known for *P. striiformis* (Schafer 1987). Chemical control of rust diseases with fungicides has been successful, but the cost of the fungicides, the economics of US wheat production, and concerns about chemicals in food grains have limited their use in the United States (Rowell 1985). Fungicides are widely used in Europe and the Pacific Northwest to control other wheat diseases.

By far the most common approach to the control of rust diseases in wheat is the use of conventionally bred resistant plants because it costs less than fungicide applications and there are numerous sources of genes for pest-protection (for example, McVey 1990; Cox et al. 1994). The most important aspect of breeding for rust-protection is that not only the genet-

TABLE 3.1   Wheat Yield Losses Due to Stem, Leaf, and Stripe Rust in
United States, 1995-1998

| | | Yield loss, % of harvested bushels | | |
| | | Common | | |
Year	Disease	Winter	Spring	Durum
1995	Stem Rust	0.01	0.01	0.0
	Leaf Rust	2.36	0.10	0.0
	Stripe Rust	0.11	0.03	0.0
1996	Stem Rust	0.23	0.00	0.0
	Leaf Rust	0.78	0.03	0.0
	Stripe Rust	0.26	0.05	0.0
1997	Stem Rust	0.00	0.02	0.0
	Leaf Rust	2.85	1.10	0.0
	Stripe Rust	0.07	0.04	0.0
1998	Stem Rust	0.09	0.03	0.0
	Leaf Rust	1.60	0.83	0.0
	Stripe Rust	0.27	0.17	0.0
Average, 1995-1998	Stem Rust	0.08	0.01	0.0
	Leaf Rust	1.90	0.52	0.0
	Stripe Rust	0.18	0.07	0.0

Source: USDA (1999g).

ics of the host but also the genetics of the pathogen must be considered.
Both are subject to change—the pathogen by mutation and sexual hybridization, the host by plant breeding. Because of changes in the pathogen,
protective genes in the host are overcome by new virulence genes in the
pathogen.

Kilpatrick (1975) used an international testing program to estimate
that the average lifetime of a gene for protection from leaf, stem, or stripe
rust was 5-6 years. The rapid loss of genetic pest-protection due to new
virulence genes led researchers to look for new protective genes and for
durable resistance (for example, Line 1995). R. Johnson (1984) has defined
durable resistance as the "resistance that remains effective during prolonged and widespread use in an environment favorable to the disease."
Considering the definition, genes for durable resistance are identified
only after they have been deployed in widely grown cultivars.

New genes for pest-protection are constantly being searched for in
wheat and its wild relatives, and plant breeders try to create new combi-

nations of protective genes. Increasingly, new genes are being identified and transferred from wild relatives of wheat (for example, Cox et al. 1993 and 1994; Sharma and Gill 1983). Wild relatives of cultivated plants have coevolved with the crop pathogens and so are often extremely useful sources of protective genes (Leppik 1970; Wahl et al. 1984). Although many of the wild relatives of wheat are *Triticum* spp., many are more distant (McIntosh et al. 1995). For example, the protective genes *Lr24* and *Sr24* came from tall wheat grass, *Thinopyrum ponticum* (Podp.) (Barkw. & Dewey); and *Lr26, Sr31,* and *Yr9* came from rye, *Secale cereale* L. Little is known about the biochemistry of genetically based rust-protection, so most breeding programs use phenotypic selection (the presence or absence of the disease) and some use molecular markers to track protective genes.

The main phases of any wheat-breeding program are introduction of genetic variation, inbreeding and selection of useful variants, and extensive field testing of selected variants to determine their agronomic or commercial worth (Baenziger and Peterson 1992). All the standard plant breeding methods are well documented (for example, Fehr 1987; Stoskopf et al. 1993), as are the methods specifically applied to breeding for rust-protection (Knott 1989; MacIntosh and Brown 1997). The most common breeding method for moving one or a few genes into an elite line or cultivar, especially when the genes are being transferred from a wild relative or an unadapted line, is backcrossing. It has been widely used to introduce protective genes into cultivated wheat whether those genes are derived from *Triticum* spp. or from more distant but sexually compatible relatives.

The Agricultural Result

As mentioned previously, the effect of rusts can be devastating when susceptible wheat cultivars are grown. However, estimating the value of crop resistance to rust accurately is difficult because the widespread growth of resistant cultivars affect the yield-loss estimates. A well-documented estimate (based on potential yield losses due to the disease-infecting susceptible cultivars) of the annual value of having stem rust resistance in wheat grown in western Canada was Can$217,000,000 (Green and Campbell 1979). The annual yield losses due to stem rust have averaged between 15% in Saskatchewan to 25% in Manitoba. In the United States, epidemics are localized, but the yield losses due to plant susceptibility to stem rust were as high as 56.5% in North Dakota and 51.6% in Minnesota in 1935 (Roelfs 1979). For leaf rust, the yield losses due to susceptibility were estimated at 50% in Georgia in 1972. For stripe rust, yield losses due to susceptibility were estimated at 30% in 1961 in Washington. The losses should be viewed as high estimates because the weather conditions, pathogen, and host susceptibility were optimal. The average annual US losses are probably more similar to those estimated by Green and Campbell for Canada (1979). The three rusts combined reduce the annual US wheat crop by about 2% (table 3.1); most of the losses are caused by leaf rust. The low level of rust losses is attributable mainly to the use of resistant cultivars.

Health and Environmental Impacts

No formal assessment of the health or environmental impact of conventionally breeding wheat for resistance to rust has been undertaken by regulatory agencies, inasmuch as the products of conventional plant breeding have generally not required their oversight. Rust-resistant wheat cultivars, regardless of the source of their resistance genes, have been widely grown and consumed in food products with no history of causing health problems.

Little is known about the biochemical basis or gene products for plant protection against rust, but these genes are likely to be similar to other genes in the large class of race-specific resistance genes isolated from other plants. The presence of pest-resistance genes can affect end-use quality by affecting the grain protein content. For example, leaf rust detrimentally affects leaves reducing their potential for nitrogen remobilization to the grain and reducing grain protein content (Cox et al. 1997). Lower grain protein content is generally considered a detrimental effect in hard wheats but a beneficial effect in soft wheats (Finney et al. 1987). Stem rust tends to reduce the flow of photosynthate and nitrogen to the grain; but because nitrogen is mobilized early in grain development, the overall result of stem rust is generally an increase in protein content in the grain, possibly including shriveled kernels.

Environmentally, the use of rust-resistant wheat cultivars has reduced the use of fungicides, but the extent of this reduction is not well documented, because effective pest-protection has been widely deployed for many years and the economics of wheat production often preclude the widespread use of fungicides that are effective against rust.

3.1.2   Bt Crops

The most widely used transgenic pest-protected plants are cultivars that express insecticidal proteins derived from the bacterium *Bacillus thuringiensis* (Bt). Cotton and corn are protected from some of their lepidopteran pests by Bt proteins in the Cry1A and Cry9C groups. The potato cultivars are protected against the Colorado potato beetle by a Cry3 Bt protein.

There is a tendency to consider Bt toxins as all biochemically similar, but the DNA sequence similarity among toxins can be less than 25% (Feitelson et al. 1992) and the biochemical properties of the more than 100 different Bt toxins vary widely.

Potato

Transgenic potato was the first Bt crop variety approved for commercial use (EPA 1995a). The target pest for transgenic Bt potatoes is the Colorado potato beetle. This pest is not a major problem in all areas of potato production, but the need for an alternative to conventional insecticides for controlling it by conventional farming techniques was apparent in the years before approval because the beetle had become resistant to all available classes of conventional insecticides. Just as Bt potatoes reached the market, a novel insecticide, imidicloprid, also reached the market. The new insecticide was so effective on a number of potato pests that it competed effectively with Bt potatoes that controlled only the beetle pest. In 1998, Bt potatoes in the United States were planted in 50,000 acres, which is 3.5% of the total US potato acreage (Idaho Statesman 1998). Strains of Colorado potato beetle resistant to imidicloprid are already evolving in a number of locations (for example, Suffolk County, NY), so Bt potatoes may soon be planted on a much larger scale. However, low rates of adoption of Bt potato may ultimately be due to the need for potato growers to use chemicals to control insect pests other than the Colorado potato beetle. In those cases, protection from the Colorado potato beetle may not offset the cost of the chemicals and the transgenic seed (Gianessi and Carpenter 1999).

Some of the Cry3 protein produced in Bt potatoes is coded from the full length bacterial gene. However, a significant fraction of the Cry3 toxin produced in Bt potato is a smaller, truncated form of molecule (Perlak et al. 1993). EPA documentation indicates that the potential for the smaller Cry3 molecule to induce a food allergy is similar to that of the larger Cry3 molecule. EPA (1995c) indicates that "despite decades of widespread use of *Bacillus thuringiensis* as a pesticide there have been no confirmed reports of immediate or delayed allergic reactions from exposure". Bt toxin's history of use (microbial Bt sprays have been registered since 1961) and rapid digestion in simulated gastric fluids (in less than 30 seconds; EPA 1995c) are considered evidence of safety by the EPA. In addition, in acute toxicity studies, no adverse effects were exhibited (Lavrik et al. 1995) (see section 3.1.3). However, it must be recognized that the microbial Bt toxins that have been widely used for decades to control lepidopteran pests have less than 60% molecular similarity to the toxin which is active against Colorado potato beetle (Feitelson et al. 1992). This beetle-active toxin has only been used in spray form since the mid-1980s on limited acreage of potatoes. Also, the sprayed beetle toxin is not applied to the tuber; whereas, in the transgenic potato varieties, it is present in the tuber (Rogan et al. 1993).

As indicated in section 2.6, field studies that compared the biodiversity of insects in fields with transgenic pest-protected potatoes and in fields with nontransgenic potatoes treated with synthetic insecticides found higher densities of above ground beneficial arthropods in the transgenic Bt fields. USDA referred to the findings in its positive response to Monsanto's request for nonregulated status of transgenic Bt potato (USDA 1995a). EPA's pesticide fact sheet for Bt potatoes (EPA 1995a) did not refer to those field data but concluded that no negative ecological effects of Bt potatoes were expected on the basis of a series of laboratory tests conducted by Monsanto (for example, Sims 1993; Keck and Sims 1993).

Details of methods and results of the laboratory tests were voluntarily provided to the committee by Monsanto. Examination of this information indicated that most of the procedures and conclusions were valid. However, in some cases, the approach to testing seemed inefficient. For example, tests for nontarget effects on honeybee larvae used a bioassay in which 5 uL of a Bt-toxin solution was pipetted into the bottom of larval cells and observations for potential mortality were made (Maggi 1993b); this approach would be better for a contact toxin than for toxins such as Bt toxin, which must be ingested. In the tests for adult honeybees (Maggi 1993a), the amount ingested was estimated by weighing the solution before and after presentation to honeybees and controlling for evaporative loss; however in the larval study (Maggi 1993b) the amount ingested was not estimated and it is unclear how much of the solution was consumed by the larvae. A positive control (that is, a group of larvae presented with a solution that will definitely kill them) was not mentioned in the study provided to the committee.

For some other tests, the absence of information made interpretation of results less clear. For example, tests with ladybird beetles used adults and provided the Bt toxin in a honey solution (Hoxter and Smith 1993). Consumption was measured by comparing the weight of the test diet before and after presentation to the beetles. However, the measurement of consumption was not useful, because there was no control for evaporative loss. Without knowledge of the amount consumed, it would be better to gather data on egg production which is more sensitive to stress. Tests of larval ladybird beetles would also offer a more sensitive toxicity test for Bt.

The ladybird beetle test and other tests for EPA used Bt toxin produced by bacteria instead of plants. In some of the nontarget testing, that seems to have been done to increase the toxin concentration to more than 100 times the concentration in pollen[1] or nectar. That is justified, but it would also be ecologically relevant to determine effects of the actual pollen and nectar produced by the plants under field conditions. In soil-degradation studies in which the bacterially produced toxin is used at concentrations that could be obtained from the plant itself, there seems to be less justification for not using the plant itself. It is surprising that in the Bt soil-degradation studies, either bacterially produced toxin or freeze-dried and highly pulverized Bt-potato plant material is used (Keck and Sims 1993). An ecologically more realistic approach was used in peer-reviewed studies by Donegan et al. (1995) and by Palm et al. (1996): where Bt and non-Bt plant material was placed in the field and monitored for decomposition, microbial diversity, and Bt-toxin titer. Donegan et al. (1996) also used an ecologically realistic system to test for differences in rhizosphere and leaf-dwelling microorganisms associated with field-grown transgenic Bt and non-Bt potatoes. No biologically significant differences were found. Similar tests would be valuable in regulatory assessments.

Overall, the data presented to EPA by Monsanto demonstrate that the transgenic Bt potatoes are likely to be environmentally much less disruptive than current chemical control practices against the Colorado potato beetle.

The concentration of toxin produced in the foliage (19.1 µg/g) of Bt potatoes (EPA 1995a) far exceeds the concentration needed to kill young Colorado potato beetles (Perlak et al. 1990 and 1993). This level of toxin is expected to fit the EPA Scientific Advisory Panel's (SAP 1998) definition of a "high dose" that can be useful in delaying evolution of resistant pest strains (see section 2.9). The concentration of toxin in the tubers themselves is low (1.01 µg/gm), but potato beetles do not typically feed on the tubers.

Corn

Unlike the Colorado potato beetle, which can devastate potato production in some areas when not controlled with insecticides, the European corn borer, which is the major target of transgenic Bt field corn, has not commonly been controlled with insecticides. A survey of the literature (Gianessi and Carpenter 1999) indicates that across the corn belt only

---

[1]Some varieties of potatoes (for example, Russet Burbank) produce very little pollen, so the discussion of pollen refers to other potato cultivars that have been commericialized (for example, Superior).

5.2% of the acreage is sprayed annually for corn borers and in Iowa only 2.6%. Some of the reasons for the lack of chemical control are that the perceived yield loss has always been considered small (estimated at about 4%), the cost of pesticides is high relative to the crop's value, and typical insecticides have not been very efficient at killing the pest after it bores into the plant.

In addition to the European corn borer, other insect pests are targeted by the Bt corn cultivars. For example, the southwestern corn borer, a major corn pest in some areas of the Midwest, can be controlled by Bt corn; and the corn earworm, a minor pest of field corn but a major pest of sweet corn, is also a potential target. A recent USDA study (1999d) indicates that in two of five regions mean yields were significantly higher with Bt corn than with non-Bt corn. How the increase in yield will affect farmers' profits is not evident, given increased seed cost and the increased potential for higher national production of corn which could lead to a decrease in prices.

At least three Bt toxins are produced by commercial transgenic pest-protected corn cultivars. The most common, Cry1Ab, is produced either as a full-length protoxin, as produced in *B. thuringiensis* (Monsanto variety), or as a truncated, preactivated toxin (Novartis variety). In nature, the 130 to 140 kilodalton protoxin is converted to a 60-65 kilodalton protein when the target insect ingests Bt (Federici 1998). Cry1Ac toxin is produced by the Dekalb cultivars; and a biochemically distinct Bt toxin, Cry9C, is produced by corn developed by AgrEvo (EPA 1998c). All the corn cultivars that produce Cry1A toxins have been approved for human consumption, but currently Cry9C corn has been approved only for cattle feed (EPA 1998c). That restriction by EPA has been established because the Cry9C toxin, unlike the Cry1A toxins, does not degrade rapidly in gastric fluids and is relatively more heat-stable; these characteristics of Cry9C raise concerns including those of allergenicity.

Novartis-produced Bt corn does not produce detectable levels of Bt toxin in the silks or corn kernels, so it does not effect the corn earworm, which feeds mostly on these two plant parts. The Cry9C toxin in AgrEvo corn is produced in the silk and corn ear, but it is not toxic to the corn earworm. The biological complexity of current Bt corn products is much greater than that of Bt potato and cotton; for example, only one company has commercialized Bt potato, and only one toxin type and seasonal tissue distribution are exhibited for each crop species.

Environmental impacts of Bt field corn must be judged against the typical corn system in which no insecticides are applied to control lepidopterans. Peer-reviewed studies have demonstrated adverse effects of Cry1Ab on predaceous lacewings (Hilbeck et al. 1998a, b), however, none of the studies conducted for EPA by the registrant has found adverse effects on lacewings or other aboveground beneficial insects (EPA 1997a and 1998a).

It is difficult to reconcile the different findings of the studies conducted for EPA by Monsanto (for Cry1Ac and Cry1Ab toxins) and the studies by Hillbeck and colleagues (Cry1Ab). In the first study by Hilbeck et al. (1998a), lacewings were fed on small larvae of Bt-sensitive and Bt-nonsensitive herbivores that had eaten vegetative-stage Bt or non-Bt corn. The concentration of toxin to which the lacewings were exposed could have been above the 50 parts per million (50 ppm) expected in an ecologically realistic system. A total of 200 lacewings were used per treatment. The second Hillbeck et al. study (1998b) fed larvae purified bacterially-produced Bt at a concentration of 100 ppm in an artificial diet. In the Monsanto studies, the concentration of toxin was 20 ppm and involved coating lepidopteran eggs with bacterially produced toxin (Hoxter and Lynn 1992). In each Monsanto study, 30 lacewings were used per treatment. The Hilbeck et al. (1998a) and Monsanto studies followed larvae to pupation. The Hilbeck studies found more than a 50% increase in mortality; the Monsanto studies found no difference in mortality or lower mortality associated with Bt treatment. Because lacewings typically feed only on the internal content of the eggs, they may not have ingested much of the toxin which was deposited on the shells of the eggs in the Monsanto study. Given that Bt corn is already planted over millions of acres in the United States, it seems appropriate for EPA, USDA, or registrants to sponsor careful field tests to determine whether lacewings or other natural enemies of crop pests are adversely affected by Bt corn. One preliminary study of this type found no differences between Bt and non-Bt corn in effects on any natural enemies of crop pests (Pilcher et al. 1997), but more detailed studies would be useful. Likewise, the committee recommends that

EPA should provide guidelines for determining the most ecologically relevant test organisms and test procedures for assessing nontarget effects in specific cropping systems.

Peer-reviewed studies (for example, MacIntosh et al. 1990) demonstrated that the Bt toxin in corn could affect many lepidopteran species. A laboratory study showed that pollen from some Bt corn cultivars can kill and slow growth of monarch caterpillar larvae if enough pollen is placed on the milkweed leaves fed to the caterpillars (Losey et al. 1999)(see section 2.6.2). If monarchs are indeed being killed in nature by this pollen, the non-Bt corn planted as a refuge for susceptible pest insects could be planted around the edges of corn fields so that adjacent milkweed would be dusted only with pollen from non-Bt corn. It might also be possible to shift to using Bt corn which does not produce biologically significant amounts of Bt in its pollen (EPA 1998b; Andow and Hutchinson 1998).

The potential for resistance to Bt toxin in a number of the target pests of Bt corn has been of concern to EPA, environmentalists, and university researchers (Ostlie et al. 1997; Andow and Hutchison 1998; Matten 1998; SAP 1998). In particular, the corn earworm may be especially vulnerable to evolving Bt resistance. Corn earworm is substantially less sensitive to Bt toxins than the primary target pest of Bt corn, the European corn borer. Bt corn varieties that express the toxin in the silks or corn kernels where corn earworm feed do not produce a high enough dose for corn earworm mortality. Corn earworm is also subject to selection pressure from Bt toxins in Bt cotton, since this pest feeds on a number of crops, including cotton, where it is known as the cotton bollworm (EPA and USDA 1999).

All the commercial cultivars provide substantial protection against the European corn borer (Ostlie et al. 1997). However, the Novartis-produced cultivars, which use green tissue and pollen-specific promoters to drive gene expression, have lower efficacy later in the season (Ostlie et al. 1997; Andow and Hutchison 1998). The lower late-season efficacy is also seen in the Dekalb-produced corn (Andow and Hutchison 1998). Lack of a high dose in these two types of Bt corn could undermine the high-dose refuge approach endorsed by EPA (Matten 1998; and section 2.9) and achievable with other Bt corn cultivars.

Cotton

Like potatoes, conventionally-grown cotton has been heavily treated with insecticides to control lepidopteran pests. Therefore, the introduction of Bt cotton can produce considerable environmental benefits. A 1998 survey indicated a general decrease in insecticide useage on Bt cotton (Mullins and Mills 1999). For example, in 66 comparisons in the Mississippi, Louisiana, and Arkansas region, the average number of insecticide sprays per field was 10.1 for non-Bt cotton and 7.9 for Bt cotton. Many of these insecticide treatments were made to control the boll weevil which is not affected by Bt. In 20 comparisons in the North Carolina, South Carolina, and Virginia region (where the boll weevil is not a pest), the average number of insecticide sprays was 3.7 for non-Bt cotton and 1.2 for Bt cotton. USDA's Economic Research Service found less clear patterns in changes in insecticide used on Bt cotton (USDA 1999d). Comparison of mean pesticide acre-treatments for 1997 showed that in only two of three regions surveyed did the adoption of Bt cotton reduce insecticide treatments normally used to control pests targeted by Bt. In one of three regions, total insecticide treatments for all other pests was higher for Bt adopters than for nonadopters (USDA 1999d). The results should be interpreted with caution however, because, as stated in the USDA study, "attributing differences in yields, pesticide use, and profits between adopters and nonadopters observed in the data solely to adoption of genetically engineered crops is nearly impossible because many other factors also affect yield and pesticide use." However, in summary the committee concludes that

The introduction of Bt cotton appears to be bringing environmental benefits through the reduced use of insecticides.

Monsanto has provided EPA with data on potential effects of the Cry1Ac toxin on nontarget species (for example, Sims 1994) and has made the information available to this committee. The range of nonpest species tested is similar to that of species tested for registrant's corn, and the test protocols were also similar. None of the studies showed biologically significant impacts on nontargets other than lepidopterans. The only ecological study in which a comparison of corn and cotton testing produced contrasting results was the study of toxin degradation in soil (Ream 1994a; Sims and Sanders 1995). In the corn study, the half-life of Cry1Ab Bt toxin from pulverized corn placed in vials of soil was 1.5 days, and 90% of the activity was lost in 15 days. In the similar study of cotton plant material that produced Cry1Ac toxin, the half-life was 41 days, and 90% of the activity was expected to be lost in 136 days (on the basis of extrapolation). These differences may reflect differences in tissue degradation perhaps resulting from the differences in woody cotton tissue compared to corn tissue. In comparison with insect-control practices now used in nontransgenic cotton, the possible minor effects of that Bt cotton may have on beneficial insects are expected to be insignificant.

The lepidopteran pests that attack cotton differ by region. In North Carolina, South Carolina, and Virginia, the cotton bollworm, *Helicoverpa zea*, is a predominant pest. In the Midsouth, the tobacco budworm, *Heliothis virescens*, is typically the most important caterpillar pest. In Arizona, the pink bollworm, *Pectinophora gossypiella*, is the only lepidopteran pest of economic importance (Gould and Tabashnik 1998). Each of those pests has a different response to the one Bt toxin, Cry1Ac, that is present in commercial cotton cultivars. The concentration of this toxin in cotton decreases toward the end of the season but still seems to be enough throughout the season to fit the EPA SAP criteria (SAP 1998) of a "high dose" for the tobacco budworm. However, this same toxin concentration range causes only 60-95% mortality in the cotton bollworm and cannot be considered a high dose for this pest. Although high mortality in the pink bollworm has been reported in the field, at least some pink bollworms survive late in the season. Those differences in Bt cotton efficacy against the cotton pest complex have caused concern for both short-term management and resistance management. From the perspective of short-term management, farmers have difficulty in distinguishing the tobacco budworm from the cotton bollworm during the egg and young caterpillar stages. In many areas, both species occur and farmers have trouble in determining when they can rely on Bt cotton for control and when they need to consider additional control tactics. That can lead to overuse of insecticides by farmers who are trying to decrease the probability of yield loss.

From the perspective of resistance management, the currently available cultivars of Bt cotton can fit well into the high-dose refuge approach to the tobacco budworm (section 2.9), but it falls short of producing adequate toxicity in the cotton bollworm. To slow the evolution of adaptation to Bt toxins by corn earworms, very large refuges of non-Bt cotton are needed (Gould and Tabashnik 1998).

The toxin titer in Bt cotton appears to be at least close to a high dose for the pink bollworm in early-season Bt cotton, but it is unlikely to achieve a high dose late in the season. Because Bt cotton has been so widely adopted by Arizona growers, refuges are on the verge of being too small, even if attainment of a high dose is assumed. Recent research indicates that pink bollworm has the genetic capacity to develop resistance to Bt cotton (Liu et al. 1999), so the overuse of Bt cotton in Arizona could lead to a rapid loss of the technology. Unlike the cotton bollworm, which is a pest on a wide variety of crops, the pink bollworm is specialized on cotton and its close taxonomic relatives. If pest resistance to Bt evolves in the pink bollworm, the problem will mostly be restricted to the cotton growers and is unlikely to have any impacts on organic or other farmers who rely on sprays of Bt bacterial formulations for the production of crops other than cotton.

3.1.3   Mammalian Toxicity Testing for Bt Crops

Only general summaries of the toxicologic assessments of Bt crops were available from EPA pesticide fact sheets. However, the Monsanto Company made available to the committee the toxicology data used in the registration process for its corn, cotton, and potatoes (see appendix B). The array of tests and the protocols used for the tests of Bt toxins from each of the three crops were generally similar. The committee therefore reviewed the three crops together. Three types of studies were conducted on each crop. The goals of the studies were

- To establish equivalency of Bt proteins produced by the three crops and Bt proteins produced by appropriate *B. thuringiensis* strains or engineered *E. coli* strains (Bartnicki et al. 1993a,b; Sammons 1994a; Lee et al. 1995).

- To test for digestibility of the microbially-produced Bt proteins by simulated gastric and intestinal fluids (Bartnicki et al. 1993b; Keck et al. 1993; Ream 1994b,c).
- To test for acute effects on mice of the microbially-produced Bt proteins (Naylor 1992, 1993a,b; Sammons 1994b).

Equivalency

Determining chemical or functional equivalence of two proteins is difficult, especially if there is any question about potential post-transcriptional modification of either protein. However, a battery of well-conceived biochemical and functional tests can provide reasonable assurance of equivalence. Tests performed on the plant and microbially-produced Bt proteins included:

- Assessment of equal migration of the full length and trypsinated forms on sodium dodecyl sulfate polyacrylamide gel electrophoresis (SDS-PAGE), and binding of the migrated proteins to a polyclonal antibody.
- Demonstration of lack of glycosylation.
- Assessment of sequence similarity of about 10-15 amino terminal amino acids and up to three short internal protein sequences.
- Determination of similar toxicities of the proteins to one or two insect species.

Protein reactivity with antibodies and protein mobility are ways to determine if two proteins are similar in structure. In SDS-PAGE immunoblotting tests of comigration, the plant- and microbially-produced Bt proteins always had similar, strong protein bands at the same gel position that reacted with the polyclonal antibody.[2] However, in many cases, there were additional bands in the gel lanes with plant-expressed Bt protein or microbially-expressed Bt protein. These additional bands differed in migration, but they bound to the polyclonal antibody. That indicates that the antibody was not highly specific or that there were different Bt proteins in the two types of preparations. Inasmuch as there was less dissimilarity in the trypsinated preparations, some of the nonhomologous bands in the full-length preparations might have been biologically unimportant. In some comparisons for cotton, a non-Bt plant-protein extrac-

---

[2]The polyclonal antibody was raised against *E. coli*-produced recombinant Bt protein.

tion was used as a control in the SDS-PAGE assessment; this offers useful information about potential binding to non-Bt proteins. A set of monoclonal (or polyclonal) antibodies with less cross-reactivity might prove useful in future tests. In general, good analytical tools to monitor Bt toxins and their concentrations are necessary for reliable human health and environmental testing.

Glycosylation appeared to be absent in all the Bt proteins on the basis of standard tests, and this result seems reasonably straightforward.

Because the DNA sequences of the Bt proteins were modified substantially in order to increase plant expression, there is always a question of whether the final amino acid sequences are identical (that is, cloning artifacts could lead to some amino acid substitutions). The sequencing of some of the terminal and internal amino acids cannot itself prove identity; however, combined with results of other tests, the finding that these sequences were identical offers additional support of equivalence.

Tests showing similar toxicity of the proteins to one or two insect species offer some support of equivalence. If such tests demonstrated substantial differences in toxicity, there would be strong evidence of nonequivalence. The finding of similar toxicity, however, does not prove equivalence. In the current studies, when two insect species were used, they were taxonomically closely related. Future tests with less related species would be useful. There is a question of how many biochemically distinct insect species should be tested in this kind of study.

Digestibility

The digestibility tests for each of the toxins were similar and reasonably straight forward. Microbially-produced Bt toxins were incubated with simulated gastric fluids and separately with simulated intestinal fluids. The gastric fluids caused the proteins to lose their biological activity against insect species and to lose their potential for binding with an antibody used for protein detection. In contrast, the intestinal fluids only truncated the full-length proteins to their active forms and did not affect the pretruncated proteins. Because the proteins reach the gastric fluids first, the lack of degradation by intestinal fluids would be unimportant unless an individual lacked active gastric digestion.

Acute Toxicity to Mice

Mice were given oral doses of the microbially-produced Bt toxins about 100-1000 times the acute dose that they would encounter in consuming one-tenth of their body weight in plant material. Ten 7-week-old mice were exposed to each dose; in no case was significant mortality, weight loss, or a general adverse clinical symptom found.

Chronic Toxicity to Catfish

There is a limit to the amount of plant material that can be consumed by mice, and that limits testing to acute tests with purified material. In a novel approach (Jackson et al. 1995), Bt-producing corn (finely ground seed) was incorporated into the diet of catfish (100 fish per treatment) over a period of 66 days. The ground corn made up 35% by weight of the catfish diet. Non-Bt corn was used as a control. No significant effects on weight gain or rates of feed conversion were found. Although this is not a traditional toxicity test for human health assessment, the duration of the test, the sample size per treatment, and the amount of plant material consumed provide useful information. Catfish are not a close physiological model for humans, but forage- and grain-consuming mammals might be more appropriate models (see section 2.5).

Overall Findings

The committee concludes that

Although a number of the experiments performed in support of registration for transgenic pest-protected plants containing Bt proteins could be improved by modifications suggested above, the total weight of evidence from combined studies presented and previous knowledge about Bt proteins, provides reasonable support for the toxicological safety of crops containing the tested Bt proteins (that is, Cry1Ab, Cry1Ac, and Cry3A).

Similar tests of other Bt proteins would be appropriate in the future, but a question remains about the strength of data that should be required when a class of plant-defensive substances has not been previously characterized as well as Bt proteins have been. As stated above, biochemical and functional equivalence is difficult to prove. Functional activity can be affected by single amino acid substitutions that would be difficult to detect with current methods. Although the sequence of the cloned gene is usually known and will largely remain intact during plant expression, modifications such as amino acid substitutions, proteolytic processing, and glycosylation are all possible. Therefore, it would be helpful to use plant-produced defensive substances in as many tests as is feasible. However, because it may often be difficult to extract and purify pest-protective proteins from the plant itself, EPA should provide guidelines for determining when a similar protein produced by another organism can be considered equivalent to the plant-produced protein and be used in toxicological and environmental testing (see recommendation in chapter 2, section 2.5.1).

Research should be conducted to assess the relevance of using forage- and grain-feeding organisms in chronic toxicity studies (see section 2.5.2). Such studies have historically been conducted in evaluating new conventionally-bred forage-crop varieties and have often detected effects of new varieties on animal performance traits (Hanson et al. 1973; Reitz and Caldwell 1974).

3.1.4   Virus-resistant Squash and Papaya

Many crops are damaged by viruses, sometimes to the extent that these pathogens limit the regions where the crop can be grown. Viruses are often transmitted by aphids and other insects that are difficult or impossible to control effectively with pesticides. Conventional breeding for virus-protection is sometimes possible, but naturally occurring protective genes are not always available in the crop or related species. Furthermore, the rapid evolution and spread of new viral strains often thwarts efforts to achieve durable protection in the crop.

Transgenic methods can provide much-needed protection from viruses by transferring pieces of the viral genome into plants (Sanford and Johnston 1985; Powell-Abel et al. 1986). In the future, it might be possible to avoid viral infections by developing plants that have transgenic protection from aphids and other vectors. Here the committee discusses the first transgenic virus-protected crops to be granted nonregulated status and sold commercially: Asgrow's crookneck squash varieties (deregulated in 1994 and 1996) and Cornell University's papaya (deregulated in 1996). Virus-protected potato has also been approved for marketing, and more than 20 other domesticated species have been field-tested to evaluate transgenic protection against viruses (chapter 1, tables 1.4 and 1.5).

Squash

Varieties of domesticated *Cucurbita pepo*—commonly known as zucchini, yellow crookneck squash (summer squash), or acorn squash—are widely cultivated in the United States for human consumption. In some regions and some years, viral infections stunt and mottle, or deform the squash, causing major economic losses and even complete failure of the crop. The outbreaks are often intermittent, however, and therefore difficult to predict (for example, Schultheis and Walters 1998). Some of the most common cucurbit viruses in the United States are watermelon mosaic virus 2 (WMV2), cucumber mosaic virus (CMV), zucchini yellow mosaic virus (ZYMV), and papaya ringspot virus (PRSV, formerly WMV1). Each of those viruses has many host species in the curcurbit family and others, and all are transmitted by aphids (Price 1940; USDA 1994b). Infections in any given area are sporadic, and by the time symptoms are visible it is often too late to control the aphids with pesticides (for example, Tricoli et al. 1995).

Protection from viruses in cucurbit crops has been improved somewhat by the use of genes from wild and cultivated species (for example, Gilbert-Albertini et al. 1993). Cucurbits are well known for their ability to hybridize (Wilson 1990), but species barriers have often made it difficult to transfer desirable genes into commercial varieties. Genetic protection from WMV2 and ZYMV was developed nearly simultaneously by both transgenic and conventional breeding. The Harris Moran company released a conventionally bred zucchini known as Tigress with protection from WMV2 and ZYMV (USDA 1994b; Schultheis and Walters 1998), and Upjohn/Asgrow used viral coat protein genes to achieve similar objectives in transgenic yellow crookneck squash. Upjohn/Asgrow's first commercial transgenic product was Freedom II, which was protected from WMV2 and ZYMV and was deregulated in 1994; it exhibits strong protection from the targeted viruses (Fuchs and Gonsalves 1995; Tricoli et al. 1995).

In a 1995 field experiment in North Carolina, both the conventional Tigress and the transgenic Freedom II exhibited better protection from viruses than most other varieties tested (Schultheis and Walters 1998). Asgrow then produced another transgenic pest-protected squash, known as CZW-3 (deregulated in 1996), with a marker gene for resistance to the antibiotic kanamycin and coat-protein genes for protection from three cucurbit viruses. Squash plants with transgenic protection from as many as five viruses have been field-tested (USDA 1999c) and might eventually be considered for deregulation.

The health concerns about transgenic virus-protected squash have been related to both viral and bacterial genes that are expressed in all the plant's cells, including the edible portions. Human or animal consumption of plants with viral coat proteins is widely considered to be safe, on the basis of common exposure to these proteins in nontransgenic types of food. However, Asgrow's 1996 virus-protected squash also had a marker gene for resistance to the antibiotic kanamycin. Some people have proposed that the widespread use of antibiotic-resistance genes as markers for transgenic traits could exacerbate current losses of antibiotic effectiveness due to overuse. However, for markers in current commercial use, that risk is extremely small (Nap et al. 1992; Fuchs et al. 1993), and the

Food and Drug Administration (FDA) has approved many crops with transgenic kanamycin resistance. FDA continues to examine the safety of antibiotic-resistance genes and issued the following draft statement in 1998 (FDA 1998):

> FDA acknowledges that the likelihood of transfer of an antibiotic resistance marker from plants to microorganisms in the gut or in the environment is remote and that such transfer, if any, would likely be insignificant when compared to transfer between microorganisms and, in most cases, would not add to existing levels of resistance in bacterial populations in any meaningful way.

Nevertheless, in the future, it may not be necessary to take even the small risks associated with the use of antibiotic resistance markers, since other types of markers could be substituted in future transgenic crops (for example, Kunkel et al. 1999).

The major environmental risks that have been discussed in connection with virus-protected crops pertain to effects of viral coat protein genes on the pathogenicity of other viruses (Falk and Bruening 1994) and consequences of crop-to-wild gene flow that could allow beneficial transgenes to move into feral-crop plants or closely related weeds, as described in section 2.7. The first issue was studied experimentally at Cornell University by Dennis Gonsalves, who helped Asgrow to develop the virus-protected squash; he and his collaborators concluded that the risks that other viruses would become transmissible (from heteroencapsidation) or that the nonpathogenic viruses would become more virulent (from recombination) were exceedingly small (Fuchs et al. 1998; also see section 2.8).

The second issue—whether wild relatives could benefit from disease-protection genes—was more controversial. An important precedent for USDA in dealing with this potential problem was established in the following example. To get more information on the ecology and systematics of weedy, free-living *Cucurbita pepo* (FLCP), the USDA Animal and Plant Health Inspection Service (APHIS) commissioned a report on the risks that may be posed by crop-to-wild gene flow by Dr. Hugh Wilson, an expert on cucurbit taxonomy and ecology. Wilson concluded that FLCP is a significant agricultural pest that might benefit ecologically from protection from ZYMV and WMV2 (Wilson 1993). Key information about squash and its weedy North American relatives is summarized below with an evaluation of USDA's conclusions.

Volunteer squash plants are not known to spread and become weeds, but *C. pepo* crosses freely with a wild weedy species that is variously known as *C. texana* (Texas gourd) or, more recently, wild or free-living *C. pepo* (FLCP), and *C. pepo* subspecies *ovifera* (same species as zucchini and crookneck squash; Decker 1988, Wilson 1993). FLCP is an agricultural weed in cotton and soybean fields, where its tough gourds interfere with harvesting machinery (Harrison et al. 1977; McCormick 1977; Oliver et al. 1983; Bridges 1992). Cultivated squash and free-living squash co-occur in many regions of Texas, Louisiana, Alabama, Mississippi, Missouri, and Arkansas (Wilson 1993). The bouyant gourds of FLCP are widely dispersed during floods, and this leads to periodic infestations of crops along river floodplains. Once FLCP becomes established in an area, it can be difficult to eradicate, because some seeds germinate throughout the growing season, other seeds remain dormant for long periods, and the plants spread laterally by branching and producing roots along the nodes of their long, viny stems. USDA asserts that FLCP is only a minor weed that can be easily controlled with herbicides (such as Cobra, bromoxynil, and glyphosate; USDA 1994b and 1996b). Nonetheless, FLCP is clearly a weed that merits close attention from regulatory agencies. It was previously listed as one of the top 10 most important weeds in Arkansas (McCormick 1977). Additional fitness-related traits such as virus-protection could potentially increase the agricultural impacts of this weed.

Hybridization between cultivated squash and FLCP is known to occur across distances of 1 km or more. Cultivated squash and wild squash have single-sex flowers that make them dependent on insect pollinators for seed set, and bees are known to carry crop pollen to wild plants as far as 1.3km (Kirkpatrick and Wilson 1988). Although cultivated squash hybridizes and backcrosses with weedy FLCP, USDA stated that "there is no scientific or anecdotal evidence that supports the contention that hybrids between yellow crookneck squash and FLCP plants are weeds and are persistent" (USDA 1994b). That statement appears to overlook the fact that after the first hybrid generation, spontaneous backcrossing with wild plants can allow crop genes to spread into FLCP populations, which are clearly persistent weeds. Crop-wild hybrids and their offspring are vigorous and fertile, so neutral or beneficial transgenes could probably persist in wild or weedy populations (Fuchs and Gonsalves 1999). Furthermore, virus-protection transgenes confer strong protection in crop-wild hybrids and backcrossed generations, as expected (Fuchs and Gonsalves 1999). The environmental question posed by this situation is "will genes for virus-protection be beneficial enough to cause this weed (FCLP) to become more common?"

Wild *C. pepo* and cultivated *C. pepo* are susceptible to the same viruses (Provvidenti et al. 1978). To check for viral infections in FLCP populations, Asgrow conducted a survey in 1993. Its analyses of 14 FLCP patches in nine locations did not detect viral infections in wild plants; one plant was sampled at each location (USDA 1994b). No information was given on whether cultivated squash in these areas were infected with ZYMV or

WMV2; that information would have shown whether the viruses were present that year. Because of a severe drought, no FLCP plants were sampled in Texas. On the basis of anecdotal reports and this one-season survey of only nine locations, USDA-APHIS concluded (USDA 1994b):

> Given the available knowledge, it is unlikely that resistance to ZYMV and WMV2 infection will confer a selective advantage or be maintained in the FLCP populations. Surveys of natural FLCP populations for the incidence and severity of ZYMV and WMV2 infections suggest that resistance to these viruses will confer little, if any, selective advantage, because disease caused by these viruses is apparently not among the factors important to the survival or reproductive success of FLCP.

The issue merits further empirical study, especially because selectively neutral crop genes *are* often maintained in the gene pools of wild and weedy plants (for example, Whitton et al. 1997). Also, the selective benefit of such genes could vary geographically and over time, and a small-scale survey like Asgrow's could easily miss infections that affect long-term population dynamics. No studies other than the one just mentioned have been conducted to determine whether viral diseases are important to the survival or reproductive success of FLCP.

Questions about the weediness of FLCP were addressed again when Asgrow requested deregulation of the CZW-3 squash in 1995. The CZW-3 squash is resistant to CMV (cucumber mosaic virus), as well as to ZYMV and WMV2. To the committee's knowledge, USDA-APHIS did not obtain any new, original data on the agroecological factors that regulate FLCP populations in their geographic range (USDA 1996b). Instead, the 1996 deregulation of the CZW-3 again relied on the 1993 Asgrow survey and anecdotal evidence from interviews with several weed scientists in Arkansas (USDA 1996b). Some of the reasoning in the permit document is not well supported. For example, USDA states that the arrival of ZYMV in the United States in the 1980s did not lead to decreases in FLCP populations, as would be expected if ZYMV suppressed FLCP populations. That statement is puzzling in light of the statement that FLCP populations apparently have become *less* of a weed problem in the 1990s, presumably because of changes in available herbicides (USDA 1996b). Without any studies of FLCP populations, one cannot rule out the possibility that viral diseases and other factors (such as frequency of floods) have also played a role in suppressing FLCP.

In summary, the committee concludes that

USDA's assumption that transgenic resistance to viruses will not affect the weediness of wild *C. pepo* might be correct, but longer-term empirical studies are needed to determine whether this is true.

In addition, the cumulative effects of additional transgenic protective traits on this weedy crop relative deserve further scrutiny, especially if the traits are linked and inherited as a unit. Protection from several common cucurbit viruses is expected to have a greater effect on weedy populations than resistance to only two or three.

In conclusion,

USDA's assessment about how the spread of virus-protective transgenes will affect free-living *C. pepo* populations is not well supported by scientific studies.

A precedent was set with the deregulation of Freedom II, which had protective traits similar to those of a conventionally bred variety. On the basis of its approval of Asgrow's CZW-3 squash, USDA seems unconcerned about incremental increases in the number of viral-protection traits that will be transferred to weedy wild relatives of squash. In cases like this, the committee recommends that

USDA should require original data to support agency decision-making concerning transgenic crops when published data are insufficient.

In cases when crucial scientific data are lacking about the potential impacts of gene flow on wild or weedy relatives, the committee recommends delaying approval of deregulation pending sufficient data (for example, surveys from several years over several regions), establishing a scientifically rigorous monitoring program in key areas to check for undesirable effects of resistance transgenes after the transgenic pest-protected plant is commercialized, or restricting the initial areas where the plants can be grown.

Restricting the areas where the squash can be initially grown would be preferable to unconditional deregulation, at least until more data are available.

Papaya

Papaya (*Carica papaya*) is an important fruit crop in lowland regions of many subtropical and tropical countries, including Brazil, India, Mexico, Thailand, Vietnam, and Australia. It is a fast-growing, tree-like herbaceous plant that bears fruit in its first year and is usually replanted after 2 years, when the fruits are too high to harvest easily. In commercial plantations, papaya is often infected by the common, aphid-transmitted papaya ringspot virus (PRSV), which causes severe stunting and low fruit yields (Gonsalves 1998). The effects of this virus are so severe that year-round cultivation might not be profitable and infected plantations are often abandoned. Efforts to control the aphids, or to conventionally breed for protection from the virus have been unsuccessful. Tolerance of the virus can sometimes be achieved by deliberately infecting the plants with a mild strain of PRSV, but this is laborious, unpopular with farmers, and only partially effective (Gonsalves 1998).

Papaya was introduced to the Hawaiian island of Oahu in the 1940s and PRSV began infecting the plantations in 1945. The industry then moved to the island of Hawaii, as did the virus; the small, isolated region of Puna remained virus-free until 1992. Meanwhile, Dennis Gonsalves and his colleagues at Cornell University began developing transgenic papaya with protection from PRSV in the hope of rescuing Hawaii's tenuous papaya industry. First they fused a coat protein gene from a mild strain of PRSV to a kanamycin-resistance marker. The linked transgenes were then inserted into the genomes of two local cultivars, christened UH Rainbow and SunUp. The cultivars were highly protected from the Hawaiian strains of the virus and were deregulated in 1996; this allowed the industry to begin recovering from its complete collapse. The virus is unlikely to disappear, however, because it also infects cucurbits and other hosts that occur nearby. Farmers expect transgenic protection to be a boon to the local economy, but the boon could be temporary if the pathogen evolves a way to circumvent the plants' protective mechanism. Another concern is the possibility that other races of PRSV will reach Hawaii. The transgenic papaya is not protected from isolates of the PRSV from outside Hawaii, but further research by the Gonsalves group suggests that protection from a broader spectrum of PRSV races can be obtained (Gonsalves 1998). Parallel research projects are now under way in other papaya-growing countries.

The human health risks that were evaluated before deregulation of transgenic papaya are similar to those described above for squash. As with Asgrow's squash, the papaya's viral coat protein is not expected to jeopardize human health, because consumers already ingest this compound in nontransgenic food. Gene flow to feral or wild relatives was not an issue because no wild relatives occur in Hawaii, Puerto Rico, or Florida, and the crop itself is not weedy.

3.2 ANALYSIS OF THE 1994 AND 1997 PROPOSED ENVIRONMENTAL PROTECTION AGENCY RULES FOR PLANT-PESTICIDES

The above case studies described the type of data and information used for regulatory review of transgenic pest-protected plants. This section describes the overarching proposed EPA rule for reviewing such plants and its scientific basis.

In 1994, EPA published a proposed rule on the regulatory status of plant-produced pesticides (EPA 1994a), stating that "the substances plants produce to protect themselves against pests and diseases are considered to be pesticides under the Federal Insecticide, Fungicide, and Rodenticide Act (FIFRA) definition of pesticide" and that "These substances", not the plants, "along with the genetic material necessary to produce them, are designated 'plant-pesticides'." "Recognizing the unique characteristics of plant-pesticides", EPA proposed to "create a new part in the [Code of Federal Regulations] for regulations unique to plant-pesticides" (see also section 1.5.3). In 1997, EPA published additional material (EPA 1997b) related to the 1994 proposed rule to comply with the Food Quality Protection Act (FQPA) (Public Law 104-170, EPA 1997b).

EPA proposed to regulate "a pesticidal substance that is produced in a living plant and the genetic material necessary for the production of the substance, where the substance is intended for use in the living plant." That definition excludes pesticidal substances such as pyrethrum and neem, that are produced by plants but are then extracted from the plants before being used although substances used in this way are subject to EPA regulation as conventional pesticides. The genetic material considered necessary for producing such a pesticidal substance includes genes that encode the substance itself (for example, genes coding for pesticidal proteins) or lead to the production of the substance (for example, genes coding for enzymes that convert compounds that are naturally present in the plant into pesticidal compounds). The necessary genes also include promoter, enhancer, and terminator sequences that regulate expression of the encoding genes.

EPA justified regulation of the genetic material, in addition to the pesticidal substance because the substance might not be present in all plant stages, such as seeds and pollen, although the capacity to produce the substance would be in the seeds and the pollen. Another justification for examining both encoding and regulatory DNA sequences is that variation in the concentration of the pesticidal substance in plant parts depends on specific characteristics of both the encoding and the regulatory sequences.

EPA initially proposed to regulate what it referred to as "inert ingredients" such as selectable markers that enabled researchers to test for the presence of introduced genes but were not necessary for production of the pesticidal substance. It later requested comment on dropping the regulation of inert ingredients (EPA 1996); therefore, selectable markers for transgenic pest-protected plants would be evaluated only by FDA, which has evaluated the use of these markers in the past (FDA 1994a).

The proposed rule casts a wide net that captures all plant-produced products that mitigate pest damage. EPA recognized that not all these pest-protected plant products required regulatory scrutiny. Therefore, a major portion of the 1994 proposed rule involved justification of exemptions of specific classes of pesticidal gene products from regulation under FIFRA and FFDCA. The rationale for the exemptions involves the balance between human health and ecological concerns and the positive contributions expected from pest-protected cultivars. The 1994 document states that "EPA finds that the plant-pesticides it is proposing to exempt have a low probability of risk and have potential benefits associated with them (for example, economic benefits to farmers and reducing the need for chemical pesticides) that outweigh any potential risks associated with them, and that the low probability of risk does not justify the cost of regulation." For most classes of exemptions, the scientific logic used by EPA is clear, for other classes it is questionable.

3.2.1   Exemption for Sexually Compatible Genes

Exemption from FIFRA and FFDCA

The largest class of plant-pesticides proposed for exemption by EPA includes all cases in which the "genetic material that encodes for a pesticidal substance or leads to the production of a pesticidal substance is derived from plants that are sexually compatible with the recipient plant and has never been derived from a source that is not sexually compatible with the recipient plant" (EPA 1994c, p. 60537). According to the 1994 document (EPA 1994c, p. 60537-60538), sources are

> [C]apable of forming a viable zygote through the fusion of two gametes and can include the use of bridging crosses and the use of wide cross breeding techniques such as surgical alteration of the plant pistil, bud pollination, mentor pollen, immunosuppressants, *in vitro* fertilization, pre- and post-pollination hormone treatments, manipulation of chromosome numbers, and embryo culture. Wide crosses, for the purpose of this exemption, also include ovary and ovule cultures.

As long as the genetic material comes from a sexually compatible plant, the plant-pesticide is exempt, regardless of whether the method of transferring genetic material uses sexual crosses or transgenic technology.

The committee recognizes the realistic limitations of overseeing the pesticidal substances in conventional pest-protected plants and, given their history of safe use, recognizes that there are practical reasons for exempting those substances. However, the committee questions the *scientific* basis used by EPA for this exemption because no strict dichotomy or new categories appear to exist between the risks to health and the environment that might be posed by conventional and transgenic pest-protected plant products (section 2.2.1).

The categorical exemption also applies to transgenic pest-protected plant products that contain transgenes from sexually compatible species, and the committee questions the scientific basis for this exemption as well, specifically because the genes and gene products can be expressed at concentrations far greater than the concentrations at which they are naturally expressed (sections 2.4.1 and 2.5.2) . Even though the risks of many transgenic pest-protected plants containing genes from sexually compatible species are expected to be low and would justify exemption, lack of experience with these products and public concern over genetic engineering suggest that a blanket exemption for them is inadvisable.

Major portions of the 1994 and 1997 documents explain the scientific rationale for the categorical exemption. The 1994 document states that "since traits can be passed through a plant population by sexual recombination, it is reasonable to predict that, in a sexually compatible population, new exposures of organisms that associate with plants in the population to the pesticidal substance are unlikely." It might be appropriate to exempt those plant protectants, but the rationale in the above statement and in other statements in the 1994 document disregards the following:

- Even for sexually compatible populations that are theoretically capable of natural cross-fertilization in the wild, there is no substantial passing of traits between populations unless the populations are in close proximity. Plants used as sources of new traits for commercial cultivars may come from small plant populations grown in remote locations or may be plants that are not commonly eaten by humans. New exposures could result if genes from such plants were used in commercial transgenic pest-protected plants.

- A body of scientific literature demonstrates that some populations of a plant species contain toxic compounds not found in other populations of the same species (Dirzo and Harper 1982) or contain these toxic compounds in much higher concentrations than other populations (Gould 1983 and 1988b). Therefore, even though these populations are sexually compatible, the transfer of genetic material from one population to another could result in novel or increased exposures of humans and other nontarget species.

- EPA is considering specifically traits that enhance the pesticidal nature of a crop. If those traits were not providing novel mechanisms of toxicity to or deterrence of some pests, plant breeders would not search for them and invest years of work moving them into commercial cultivars. Thus, there is reason to expect that organisms in US agroecosystems and humans could be exposed to new toxins when they associate with or eat these plants.

Plants with Pest-Protection due to Structural Transgenes from Sexually Compatible Relatives

In addition to exempting a plant-pesticide when all the genetic sequences associated with its production are from a sexually compatible plant, the 1994 proposed regulations indicate that as long as the coding sequence (that is, structural gene) is from a sexually compatible plant, the "regulatory regions and non-coding, nonexpressed nucleotide sequences may be derived from any source" and still merit a categorical exemption. Therefore, transgenic pest-protectants under the control of promoters that lead to transgene overexpression would be exempt if the structural gene for the protectant is derived from a sexually compatible species. Exempting plant-pesticides developed in this manner on the basis of an assumption of no new exposure seems to be flawed. Many coding sequences in plants are naturally expressed at very low concentrations or only in specific plant parts. If under normal conditions a plant protectant were produced only in the roots of a corn plant because of the specificity of the natural promoter sequence, a new exposure would occur if the coding sequence were spliced to a constitutive promoter that caused the plant protectant to be produced throughout the plant. Even if the plant protectant were naturally expressed throughout the plant, use of a novel promoter could increase its concentration dramatically.

The 1997 EPA *Federal Register* document addresses plant-pesticides derived from sexually compatible plants in terms of FQPA requirements (EPA 1997b). The focus is therefore mainly on human health considerations associated with FFDCA tolerance requirements. The document states (EPA 1997b, p. 27136) that

> EPA has extensively evaluated whether quantitative changes in levels of the pesticidal substances that are the subject of the proposed exemption would warrant regulation by the setting of a food tolerance. EPA has determined that changes in the levels of these pesticidal substances present a reasonable certainty of causing no harm because the highest levels likely to be attained in plants are not likely to result in overall significantly different dietary exposure.

The exemption includes all genes derived from sexually compatible plants and promoter sequences from any origin. This statement by EPA is based on the following findings stated in the 1997 document:

1) That there are few documented cases of new plant cultivars causing food safety problems despite the large numbers of new varieties introduced into commerce each year, is a reflection of the effectiveness of this process (of conventional breeding with sexually compatible plants).
(EPA 1997b, p. 27135)

2) Because knowledge of human consumption of food derived from sexually compatible plants was available and adequately addressed the issues of hazard and exposure, the Agency did not use, for the proposed exemption (59 FR 60535), data gathered in the laboratory through animal testing.
(EPA 1997b, p. 27137)

3) They (the exempted pesticidal substances) are part of the metabolic cycles of these plants. They are thus subject to the processes of degradation and decay that all organic matter undergoes. They are unlikely to persist in the environment or bioaccumulate in the tissues of living organisms. Because they do not persist, the potential for new exposures to the residues to occur, beyond direct physical exposures to the plant, would be limited.
(EPA 1997b, p. 27135)

4) The amount of pesticidal substance produced by plants normally varies among members of a closely related population (even within a single variety), because of the effects of conditions such as genetic constitution and environment (for example, weather) on trait expression. This variation in turn leads to differences in the levels and types of exposure to the pesticidal substance. Since such variation is a natural phenomenon common to all plants, humans have been and always are exposed to varying levels of the pesticidal substances that are subject of this exemption when they consume food from plants.
(EPA 1997b, p. 27135-27136)

5) Greatly increased levels of a pesticidal substance would, in general, only be accomplished at the expense of expressing other agriculturally desirable traits (for example, yield). EPA does not believe that levels of pesticidal substances that are the subject of the proposed exemption (59 Fed. Reg. 60535) will be increased to a point that will result in an adverse dietary effect.
(EPA 1997b, p.27136)

6) There is no evidence that such pesticidal substances, as a compo-

nent of food, present a different level of dietary risk for infants and children than they would for the adult population.
(EPA 1997b, p. 27136)

7)   EPA is not aware of any other substances outside of the food supply that may have a common mechanism of toxicity with the residues of the pesticidal substances that are the subject of the proposed exemption (59 FR 60535), although it cannot rule out the possibility.
(EPA 1997b, p. 27138)

8)   [T]he potential for causing adverse health effects may be more circumscribed than for traditional pesticides because, in many cases, the only significant route of human exposure may be oral.
(EPA 1997b, p. 27136)

The committee questions the scientific basis of the categorical exemption of plant-pesticides from sexually compatible plants and EPA's rationale. Although the committee agrees that there are few documented cases of new plant cultivars causing food safety problems (point 1), the committee does not believe that this provides a scientific basis for a categorical exemption of plant-pesticides from sexually compatible plants in light of the examples provided in this report. EPA's third point is questioned on the basis of evidence of indirect effects on nontarget organisms and data on the persistence of some naturally-occurring plant secondary compounds (see section 2.6). EPA's points four and five are questioned because transgenic methods can create a situation where a gene product is not regulated by the normal regulatory systems in the plant (for example, use of constitutive promoters). Additionally, information in chapter 2 indicates that there is not sufficient data on chronic effects on humans (point 2), and that some of these compounds (for example, alkaloids) share a similar mechanism of activity as do organophosphates (point 7).

Although the same scientific arguments can be made for the risks posed by conventional pest-protected plants, which are not subject to regulation under the coordinated framework, lack of experience with transgenic pest-protected products and public concern with these products constitute practical reasons for not granting a categorical exemption to transgenic pest-protectants derived from sexually compatible species. In summary, the committee recommends that

Given that transfer and manipulation of genes between sexually compatible plants could potentially result in adverse effects in some cases (for example, modulation of a pathway that increases the concentration

of a toxicant) and given the public controversy regarding transgenic products, EPA should reconsider its *categorical* exemption of *transgenic* pest-protectants derived from sexually compatible plants.

3.2.2   Exemption of Viral Coat Proteins

In addition to exempting plant-pesticides derived from sexually compatible plants, the 1994 and 1997 EPA documents propose a number of more specific exemptions. EPA generally provides more reasonable scientific justification for these exemptions. One specific class of plant products that was proposed for categorical exemption was viral coat proteins (VCPs). VCPs are already present in foods because of natural virus infections of crops and have not caused obvious medical problems, so health concerns are considered minimal. The EPA exemption of VCPs is also based on considerations that "include the low potential for adverse effects to nontarget organisms and the potential benefits (environmental and economic) of utilizing VCP (virus coat protein) mediated resistance." The committee, in general, agrees with this assessment of the minimal health and nontarget effects posed by VCP expression in crop plants (see also section 3.1.4) and concludes that

Viral coat proteins in transgenic pest-protected plants are not expected to jeopardize human health because consumers already ingest these compounds in nontransgenic food. However, the committee questions the categorical exemption of all viral coat proteins under FIFRA due to concerns about outcrossing with weedy relatives.

Although ecological concerns are discussed and a more restrictive exemption that considers outcrossing is presented, the proposed rule favors complete exemption of VCPs.

EPA should not categorically exempt viral coat proteins from regulation under FIFRA. Rather, EPA should adopt an approach, such as the agency's alternative proposal (as stated below in Option 2), that allows the agency to consider the gene transfer risks associated with the introduction of viral coat proteins to plants.

> Option 2: Exemption of coat proteins form plant viruses produced in plant with low potential for outcrossing to wild relatives. Under this exemption the Agency would limit its exemption of VCP-mediated resistance coat proteins to those viral coat protein/plant combinations that would have the least potential to confer selective advantage on free-living wild relatives.

In sections 2.7 and 3.1.4, the committee explains why the more restrictive exemption should be considered.

3.2.3   Exemption for Nontoxic Modes of Action

The 1994 EPA document requested comments on a proposal to exempt plant-pesticides that acted primarily by affecting plants and "that act through nontoxic modes of action." The types of substances that clearly are in this category are structural barriers such as plant hairs; substances that inactivate or resist toxins that are produced by pests; and substances that decrease chemical components needed for pest growth. As discussed in chapter 2 (sections 2.4 and 2.5), these exemptions are unlikely to result in any new human exposure to harmful substances.

However, within the same category the 1994 EPA document also discusses exempting plant hormones. Plant hormones often cause multiple changes in plants, including changes in secondary metabolites that might be toxic, so the scientific basis of such an exemption is questionable.

As with the exemption of VCPs, the categorical exemption of substances that act through nontoxic modes of action mostly considers human health effects. As outlined in previous sections of our report (sections 2.6 and 2.7) there is a need to consider separately the impact of such substances on nontarget species and the potential for the genes that code for these substances to move to feral populations or weedy relatives of the crop, where they could increase recipient plants' fitness. Categorical exemption under FIFRA might not be scientifically justifiable.

3.2.4   Oversight for Pleiotropic Effects

The 1994 EPA document states that

> any food safety questions beyond those associated with the plant-pesticide, such as those involving changes to food quality or raised by unexpected or unintended compositional changes, are under FDA's jurisdiction. Similarly, food safety issues associated with alterations in levels of a substance with pesticidal properties, or the appearance of a substance with pesticidal properties, that occur as an unintended consequence of modifications to a non-pesticidal trait would also fall under FDA's authority.

That is an important statement and shifts an important component of pest-protected plant assessment to FDA.

As discussed previously in this report (sections 2.4.1 and 2.5.2), genetic changes that result in production of a specific plant protectant can result in production of biologically active compounds other than the intended plant protectants. Such pleiotropic effects are sometimes difficult to predict. Furthermore, as outlined in previous sections (2.4.2 and 2.5), many approaches to producing plant protectants through the use of plants that are sexually compatible with the crop plant can result in crops that produce new compounds owing to linkage between the genes for the plant protectants and genes for the other compounds. FDA needs to address these "unintended compositional changes" carefully during their consultation process with the plant producers. USDA and EPA should also be aware of those unintended changes in evaluating the potential agricultural and ecological effects of pest-protected plants.

The committee recommends that

EPA, FDA, and USDA collaborate on the establishment of a database on natural plant compounds of potential dietary or other toxicologic concern.

The database would be publicly available and updated regularly. The following guidelines should be considered: initial emphasis should be on obtaining baseline profiles for food plants that are known to have toxic constituents and on the commonest varieties; differences among varieties, developmental stages, tissues and environmental conditions are important and should be analyzed after initial average baselines have been established; only information based on state-of-the-art chemistry and analytic methods should be incorporated; and potential information should be peer-reviewed by a committee of experts before it is added to the database (see also section 3.4.1).

3.3 SUGGESTED QUESTIONS FOR OVERSIGHT

Given the above concerns with the scientific basis of proposed oversight, the committee proposes that federal agencies use the following questions as a guide in developing their review process. These decision keys leave sufficient room for agency judgment case by case. For the most part, the agencies are following a similar logic in their decision-making, but there are some points where current decision-making does not agree with the following questions; these discrepancies are pointed out in the text.

Because the Coordinated Framework for the Regulation of Biotechnology was designed for transgenic products (see chapter 1) and the agencies do not actively assess conventional pest-protected plant products, the following questions focus on transgenic pest-protected plant products. However, the questions could be adapted and applied to nonregulatory safety assessments of conventional pest-protected plants, as the underlying concerns are not dependent on the method used to produce the plant (section 2.2.1).

3.3.1   Health Concerns: Guiding Principles

The principles in the following questions could be used to determine when a detailed analysis of health risks is warranted for transgenic pest-protected plants.

1)   Is the substance found in plant parts that consumers[3] eat or workers come into contact with?
    a)   Yes or Unknown—go to 2.
    b)   No—exempt from health concerns.

2)   Is the substance known to have general chemical and physical properties common to many allergens?
    Note: Criteria outlined in figure 2-1 could offer components for this type of evaluation.
    a)   Yes or Unknown—subject to safety assessment.
    b)   No—go to 3.

3)   Is the substance similar to substances that people now eat or come into contact with, and can confident predictions of safety based on the similarities be made?
    a)   Yes—go to 4.
    b)   No or Unknown—subject to safety assessment.

4)   Is the expected exposure to the substance substantially greater than current exposures?
    a)   Yes or Unknown—subject to safety assessment.
    b)   No—go to 5.

5)   Is there a reasonable chance, based on known properties of the substances, that its production will lead to harmful concentrations of toxicants or allergens that consumers eat or workers come into contact with?[4]
    a)   Yes or Unknown—subject to safety assessment.
    b)   No—exempt from health concerns.

---

[3]Including human and non-human consumers, such as food animals or pets.
[4]Pleiotropic effects.

EPA exempts from FFDCA and FIFRA pesticidal substances in transgenic pest-protected plants that are derived from transgenes from sexually compatible species. The committee's questions are not in accordance with that categorical exemption. Given that transfer and manipulation of genes between sexually compatible plants could potentially result in adverse effects (for example, modulation of a pathway increases the concentration of a toxicant), the categorical exemption of pest protectants solely on the basis of derivation from sexually compatible plants could be scientifically unsound in some cases.

FDA's policy for foods derived from new plant varieties is designed to address questions 1 through 5 with respect to dietary exposure to substances that are not regulated by EPA as pesticides. For pesticidal substances, EPA may consult with FDA on allergenicity issues (see chapter 4).

3.3.2   Ecological Concerns: Guiding Principles

Nontarget effects and hybridization with weedy relatives are subjects of concern for transgenic pest-protected plants. The committee suggests that a particular pest-protected plant needs to be exempt from both of these ecological concerns in order to avoid safety assessments.

Nontarget Effects: Guiding Principle

Nontarget effects are often unknown or difficult to predict. Along with standard screens for toxicity to nontarget species, comparison with agricultural practices that would occur if the transgenic pest-protected plant were not used could be made. For example, nontarget effects of transgenic Bt cotton could be compared with nontarget effects from nontransgenic cotton and the accompanying pesticide use needed to compensate for the lack of the transgenic trait. Broader environmental consequences such as changes in soil quality, wildlife habitat, or the use of fertilizers or water could be used to determine the contribution of the new variety to the sustainability of the agricultural system in which it is grown (Cook 1999). Such general environmental considerations could have effects on nontarget organisms.

However, it is important to point out that there is disagreement among scientists, including within the committee, as to whether comparison to currently used pest control practices should be the *determining* factor for allowing commercialization of a transgenic pest-protected plant. Most agree that it is one of many important factors. Therefore, both toxicity testing and field tests comparing agricultural methods are suggested. The committee recognizes that the question below leaves much room for agency judgment.

1)   Is it reasonable to expect that commercialization of plants with the transgenic resistance trait will have more substantial adverse effects on nontarget organisms than current pest control[5] has on these organisms?
   a)   Yes or More data needed to make a determination—subject to nontarget considerations.
   b)   No—exempt from nontarget considerations.

Hybridization with Wild or Weedy Relatives: Guiding Principles

The following guiding principles regarding hybridization with wild or weedy relatives are suggested for reviewing transgenic pest-protected plants. These guidelines are designed for annual crop plants and may require modification in order to address perennials. EPA's categorical exemptions of transgenic plants that have sexually compatible, nontoxic, and viral coat proteins are not in agreement with these principles in some cases. USDA analyzes these concerns according to risks posed to agriculture, so weedy relatives with agricultural effects are of concern; its methods are similar to the following questions, although original data are not always used. FDA does not provide oversight for ecological concerns.

1)   Does the cultivated plant occur in feral populations or hybridize with related species in the United States?[6]
   a)   Yes or More data needed—go to 2.
   b)   No—exempt from weedy-relative considerations.

2)   Have feral populations or wild relatives been reported as weedy or invasive in the United States or have a reasonable potential to become weedy?[7]
   a)   Yes or More data needed—go to 3.
   b)   No—exempt from weedy-relative considerations.

---

[5]Current pest control methods could include both the use of chemical insecticides or other non-chemically based methods.

[6]*Hybridization* refers to any naturally occurring gene flow that results in permanent introgression of genes from cultivated plants into noncultivated populations. Annual crops that persist for 1 or 2 years as volunteers are not considered to be feral populations.

[7]Applies to plants in both managed and unmanaged habitats. A species does not have to be included on the Federal Noxious Weed List to qualify as weedy or invasive, but it should be mentioned in peer-reviewed journals or other professional publications.

3) Does the gene for resistance confer a specific type of resistance or a greatly enhanced degree of resistance that is not found in feral populations or sexually compatible wild relatives in the United States?[8]
    a) Yes or More data needed—go to 4.
    b) No—exempt from weedy-relative considerations.

4) Is it reasonable to expect that this trait could have a substantial impact on the population dynamics of feral plants or wild relatives and will lead to increased abundance?[9]
    a) Yes or More data needed—subject to weedy-relative considerations.
    b) No—Exempt from weedy-relative considerations.

In addition to the recommendations in section 3.1.4, the committee recommends that

USDA should research, publicize, and periodically revise lists of plant species with feral populations or wild relatives in the United States in order to evaluate the impacts of outcrossing.

3.4   RESEARCH NEEDS

The committee realizes that there remain some uncertainties regarding the use of pest-protected plants, including transgenic pest-protected plants. These uncertainties can lead to ambiguities in regulation and often force agencies to base their decisions on minimal data sets. Additional research should continue to refine and improve the risk assessment methods and procedures and continue to develop additional data on both conventional and transgenic pest-protected plant products. Research along the following lines should be given priority to aid in decision-making. These categories have been chosen on the basis of the discussions in chapter 2 and this chapter. Many of these research needs are also highlighted in the executive summary (section ES.5).

---

[8]The frequency of the resistance trait might vary among populations. If the resistance trait is regarded as rare, go to 4. Also, go to 4 if the resistance trait is found only in geographically isolated populations.
[9]This will require agency judgment.

3.4.1   Health Effects Research

Methods for more efficiently and accurately identifying potential food allergens in transgenic pest-protected plants should be developed. Criteria of digestibility and overall homology with known allergens can be good indicators of allergenicity (Metcalfe 1996a), but the identification of specific protein sequences (or epitopes) involved in allergic responses, the further development of tests with human immune-system endpoints, and the development of more-reliable animal models should be pursued (section 2.5.1).

The committee suggests the establishment of a database on natural plant-defensive compounds of potential dietary or other toxicologic concern. Information needed for this database includes a clear list of what plants are used, phenotypic variation in the substances in different parts of plants, and genetic variations in different varieties. Research is needed to determine the baseline concentrations of secondary compounds in plant species of potential dietary or other toxicological concern and to determine how these compounds may vary depending on the genetic background and environmental conditions (see section 2.5.2 and recommendations in section 3.2.4).

For longterm toxicity testing, research should be conducted to examine whether longterm feeding of transgenic pest-protected plants to animals whose natural diets consist of large quantities and the type of plant material being tested (for example, grain or forage crops fed to livestock) could be a useful method for assessing potential human health impacts (see section 2.5.1).

3.4.2   Plant Breeding and Molecular Biology Research

Research on the mechanisms of pest-protection in both conventional and transgenic pest-protected plants should be encouraged so that we can produce crops that are only minimally affected by diseases and pests, deploy pest-protection strategies that have only minimal impact on the environment, and produce crops that can be consumed or used safely by humans and animals.

A major goal of current and future development of conventional and transgenic pest-protected plants should be to decrease the potential for ecological and health problems associated with some types of pest-protected plants (section 2.2.1). That includes developing breeding approaches and assays for avoiding the development of varieties with unintended high concentrations of potential toxins or decreased concentrations of essential nutrients, controlling expression of transgenes that have potential adverse nontarget effects to only nonedible plant tissues, and eliminating expression of transgenes that encode resistance factors in pollen. In addition, development of strategies that enhance the effective life span, or durability, of transgenic pest-protection mechanisms is vital. Research to develop better promoters that restrict expression of transgenes to non-edible plant tissues could lead to decreased potential for food safety problems with some pest-protected plants. Research could also lead to the more efficient use of non-constitutive promoters that result in more durable pest-protection or environmental safety. Transgenic or other techniques to decrease the potential for the spread of transgenes into wild populations should be explored.

For conventional pest-protected plants and for transgenes moved by breeding to new cultivars, the linkage of pest-protection traits to other traits carried inadvertently by the breeding process should be investigated for commercial cultivars, and more research should be conducted on potential health and ecological impacts of such linkage (section 2.4.2). Recent advances in plant genomics should help to identify the biochemical and physiological function of linked genes. Similarly, research is needed to better understand potential pleiotropic effects of pest-protection genes.

3.4.3   Ecological Research

Research to increase our understanding of the population biology, genetics, and community ecology of the target pests should be conducted, so that more ecologically and evolutionarily sustainable approaches to pest management with pest-protected plants can be developed (section 2.6). Knowledge of pests' roles in the larger biological community (for example, their role as food sources for nontarget organisms or their roles as predators of other agriculturally relevant pests) will allow us to anticipate better the indirect effects of declines in the pests due to both conventional and transgenic pest-protected plants. Knowledge of the pest population biology will enable prediction of the types of pest-protection mechanisms that would most efficiently reduce a target organism's pest status (Kennedy et al. 1987) and would help us to design more accurate resistance management plans (Gould 1998).

Research to assess gene flow and its potential consequences should be conducted (section 2.7). A list of plants with wild or weedy relatives in the United States should be established in an accessible public database (see section 3.3). This database should include the geographic locations of these relatives and could be used to determine which crop-weed complexes should be regulated. For weed species of concern (plants that might hybridize with transgenic pest-protected plants), more ecological and agricultural research is needed on the following: weed distribution and abundance (past and present), key factors that regulate weed population dynamics in managed and unmanaged areas, the likely impact of specific, novel resistance traits on weed abundance in managed and unmanaged areas, and rates at which resistance genes from the crop would be likely to spread among weed populations.

Because it is sometimes difficult to predict ecosystem level effects from small scale laboratory and field tests, longterm monitoring of pest-protected crops should be conducted after commercialization of these crops. EPA and USDA's Agriculture Research Service and Animal Health Plant and Inspection Service should encourage long-term monitoring for ecological impacts. Also, more rigorous field comparisons should be conducted to determine the relative impacts of conventional and transgenic pest-protected crops compared to impacts of standard and alternative agricultural practices on nontarget organisms.

Further studies are needed to determine the distances and densities of biologically active Bt corn pollen in the vicinity of a crop. More information is needed about the timing of pollen release, the types of insect species that would be harmed by ingesting pollen at observed concentrations, and the magnitude of mortality due to pollen versus other factors that limit nontarget populations.

3.5 RECOMMENDATIONS

• **EPA should provide guidelines for determining the most ecologically relevant test organisms and test procedures for assessing nontarget effects in specific cropping systems.**

• **The USDA should require original data to support agency decision-making concerning transgenic crops when published data are insufficient.**

• **In cases when crucial scientific data are lacking about the potential impacts of gene flow on wild or weedy relatives (for example, squash case study), the committee recommends delaying approval of deregulation pending sufficient data (for example, surveys from several years in several regions), establishing a scientifically rigorous monitoring program in key areas to check for undesirable effects of resistance transgenes after the transgenic pest-protected plant is commercialized, or restricting the initial areas where the plants can be grown.**

• **USDA should research, publicize, and periodically revise lists of plant species with feral populations or wild relatives in the United States in order to evaluate the impacts of outcrossing.**

• The EPA, FDA, and USDA should collaborate on the establishment of a database for natural plant defensive compounds of potential dietary or other toxicological concern.

• Given that transfer and manipulation of genes between sexually compatible plants could potentially result in adverse effects in some cases (for example, modulation of a pathway that increases the concentration of a toxicant), and given public controversy regarding transgenic products, EPA should reconsider its *categorical* exemption of *transgenic* pest-protectants derived from sexually compatible plants.

• EPA should not categorically exempt viral coat proteins from regulation under FIFRA. Rather, EPA should adopt an approach, such as the agency's alternative proposal, that allows the agency to consider the gene transfer risks associated with the introduction of viral coat proteins to plants.

• EPA should review exemptions of transgenic pest-protected plant products to ensure that they are consistent with the scientific principles elucidated in this report.

4

Strengths and Weaknesses of the Current Regulatory Framework

4.1 OVERVIEW OF THE REGULATION OF PLANT PRODUCTS UNDER THE COORDINATED FRAMEWORK

The executive branch formally announced its biotechnology policy on June 26, 1986, in the form of the Coordinated Framework for Regulation of Biotechnology (OSTP 1986), as reviewed in chapter 1 and described in more detail in this chapter. The three lead agencies with responsibility for implementation of the policy were the US Department of Agriculture (USDA), the Department of Health and Human Services (DHHS), and the Environmental Protection Agency (EPA). Since announcement of the coordinated framework, federal regulators have cleared the way for hundreds of new agricultural, health care, and industrial products, including dozens of plants modified through modern biotechnology.

The coordinated framework established the basis for regulation of new plant varieties produced by rDNA techniques. Although the term genetically modified is commonly used to describe these transgenic plants, it could just as easily be applied to products and plants that result from conventional plant breeding techniques (section ES.3.2) because these techniques also result in the modification of the plant's genetic makeup. The coordinated framework successfully resolved early disputes among the agencies concerning products that fall within the jurisdiction of more than one agency. For example, USDA would regulate plants grown to produce food or feed, and the Food and Drug Administration (FDA) within DHHS would have jurisdiction over the food or feed itself.

What the framework left unresolved were jurisdictional issues that would have to be addressed before commercial introduction of a number of products, including transgenic plants that were modified to resist disease and ward off insect pests. In fact, plants modified to exhibit pesticidal traits were not specifically addressed by the coordinated framework. Although it contained an extensive discussion of EPA's authority to regulate pesticides, the framework concentrated almost exclusively on microorganisms that were produced with pesticidal intent (OSTP 1986, p. 23319); this was undoubtedly because research involving transgenic pest-protected plants was at a relatively early stage.

In the 14 years since introduction of the coordinated framework, the lead agencies have worked to coordinate their oversight responsibilities and have resolved many of the issues that were either unforeseen or unaddressed in 1986. Hundreds of new plant varieties have been the subject of federally approved field tests, and dozens of new plant products are on the market today (section 1.5.5). These commercially available transgenic crops include corn, cotton, potato, squash, and papaya that are protected against harmful insects or viruses; and corn, cotton, canola, soybeans, and sugar beet that are modified to tolerate the application of herbicides. Determining which agencies have responsibility for a particular plant-related product depends on two factors: the traits that have been engineered into the plant and the use of the crops that will be harvested. A summary of the key regulatory schemes will help to put this in perspective. In general, the committee found that

Under the coordinated framework, transgenic products are subject to regulation under existing statutory authorities and USDA, FDA, and EPA are exercising regulatory oversight on that basis.

4.1.1 US Department of Agriculture and the Regulation of Plants

USDA has responsibility for protecting plants and for safeguarding American agriculture. The Federal Plant Pest Act (FPPA) provides USDA with the authority to regulate the movement into or within the United States of organisms that may pose a threat to agriculture and to prevent the introduction, dissemination ,or establishment of such organisms (US Congress 1957).[1] The plant pest definition under FPPA is listed in chapter 1, section 1.4.2 (US Congress 1957, section 150 aa(c)).

The FPPA establishes a permit system that has been expanded by USDA into a comprehensive prerelease review system for potential plant

---

[1]The FPPA supplements and extends the much older Plant Quarantine Act.

pests. Building on that system, which has been in effect for many years, USDA issued rules in 1987 designed specifically to regulate genetically modified organisms before their release into the environment or movement in commerce (USDA 1987). Those rules prohibited the introduction of so-called regulated articles without a permit from the USDA Animal and Plant Health Inspection Service (APHIS). The process typically has been used to address small-scale field testing of genetically modified plants before commercialization, and it now requires either a permit for or advance notification of the test.

Under the USDA rules, a permit is required for (1) any organism altered or produced through genetic engineering if the donor or recipient organism either (a) belongs to a group of plant pests listed in 7 C.F.R. § 340.2 or (b) is an unclassified organism and/or an organism whose classification is unknown, (2) any product that contains a listed plant pest or unknown/unclassified organism, or (3) any other organism or product altered or produced through genetic engineering that USDA determines to be or has reason to believe is a plant pest (as defined by 7 C.F.R. § 340.1). The rules define genetic engineering as genetic modification of organisms by rDNA techniques. The rules do not regulate research with genetically modified organisms in a laboratory or contained greenhouse but come into play only when a person seeks to introduce genetically modified organisms into the environment or interstate commerce.

USDA has issued some 887 permits for genetically modified organisms since the program began in 1987, primarily for limited field tests involving crop plants (USDA 1999f).[2] On the basis of its experience with the permit program, USDA has provided a number of exemptions for articles that it has determined do not pose a plant pest risk. One of the more important exemptions authorizes the introduction of certain regulated articles without a permit provided that USDA is notified in advance. To qualify for the notification process, a regulated article must be one of the plant species identified in the rule and must meet six eligibility criteria (for example, introduced genetic material must not cause the introduction of an infectious entity) and six performance standards (for example, field trials must be conducted so that regulated articles will not persist in the environment) (USDA 1987, section 3b). In the notification process, USDA must either acknowledge that notification is appropriate for the designated introduction activity (import, interstate movement, or environmental release) or deny permission for introduction and require a permit (USDA 1987, section 3e). USDA has acknowledged approximately 4,400 notifications for field tests to date; another 260 have been denied,

---

[2]Since the program began, approximately 120 permit applications have been withdrawn.

withdrawn or otherwise voided. As noted in chapter 1, about 40% of permits and notifications involve transgenic pest-protected plants.

Another important exemption allows researchers to petition USDA for a determination that an article should not be regulated as a plant pest. The rules contain detailed requirements for the data and information to be included in a petition for determination of "nonregulated status". USDA will publish a notice in the *Federal Register* and provide for a 60-day public-comment period for each petition that meets the rules' eligibility criteria. USDA has approved 50 of 69 petitions submitted for nonregulated status; the other 19 were withdrawn or found to be incomplete or void.

Before issuing a permit for the release of a regulated article into the environment, USDA must follow the requirements of the National Environmental Policy Act (NEPA; US Congress 1969) by preparing a publicly available environmental assessment and if necessary, an environmental impact statement (USDA 1995b). Before acknowledging the appropriateness of a notification or issuing a permit for an environmental release, USDA must coordinate with the state where the release is planned, submitting a copy of the application or notification to the state department of agriculture for review (USDA 1987, sections 3e and 4b).

4.1.2 The Food and Drug Administration and the Regulation of Food

The Federal Food, Drug and Cosmetic Act (FFDCA) provides FDA with broad regulatory authority over foods and food ingredients (US Congress 1958). No particular statutory provision or regulation deals expressly with food produced by biotechnology. FDA's formal position concerning such foods, as expressed in the coordinated framework, is that the statute provides ample tools for the agency to apply to meet the challenges of novel foods and biotechnology (OSTP 1986, p. 23309). That position was confirmed in 1992 on publication of a comprehensive policy statement for foods derived from new plant varieties (FDA 1992).[3]

The 1992 policy provides that foods developed through genetic modification are not inherently dangerous and, except in rare cases, should not require extraordinary premarket testing and regulation. The policy holds that genetically modified foods should be regulated as ordinary foods are unless they contain substances or demonstrate attributes that are not usual for the product. According to FDA, most food-related issues concerning

---

[3]The FDA's current policy on the labeling of foods derived from new plant varieties is discussed in the 1992 notice, 57 Fed. Reg. at 22991, and in a separate notice published in 1993, 58 Fed. Reg. 25837.

products of biotechnology will involve the application of sections 402(a)(1) or 409 of FFDCA (see US Congress 1958, sections 342(a)(1) and 348, respectively).

Section 402(a)(1) does not subject new food products to premarket approval but does establish a safety standard that can come into play depending on the circumstances presented by a given food or food constituent. The section is FDA's primary enforcement tool for regulating the safety of whole foods, including foods derived from genetically modified plants. Any person who introduces food into interstate commerce is responsible for ensuring that the food does not run afoul of the provisions of section 402(a)(1). Under FFDCA, FDA is authorized to seize adulterated food, enjoin its distribution, and prosecute persons responsible for its distribution (US Congress 1958, sections 332-334).

Under the safety standard of section 402(a)(1), food is considered to be adulterated if it contains any substance that occurs unexpectedly in food at a level that may be "injurious to health". Those substances include naturally occurring toxicants whose levels are unintentionally increased by genetic modification and unexpected toxicants that appear in the food for the first time. The policy provides guidance to the food industry in the form of flowcharts and other instructions regarding scientific approaches to evaluating the safety of foods derived from new plant varieties, including the safety of added substances that are subject to section 402(a)(1). Perhaps most important, FDA encourages voluntary consultations between producers and agency scientists to discuss relevant safety concerns.

Section 409 of FFDCA provides for the regulation of "food additives", defined broadly as including any substance "the intended use of which results or may reasonably be expected to result, directly or indirectly, in its becoming a component of food...and which is not generally recognized as safe" for such use (US Congress 1958, section 321(s)). A food additive must be approved by FDA before being used in food. The statutory mechanism for securing agency approval is the submission of a food additive petition, which must contain data and information that show a reasonable certainty that the additive will be safe for its intended use. The petition is subject to public notice and comment.

The 1992 policy acknowledges that, in some cases, whole foods derived from new plant varieties, including plants developed by new genetic techniques, might fall within the scope of section 409. It is the transferred genetic material and the intended expression product of that material in the plant that could be subject to food additive regulation if such material or expression product is not generally recognized as safe (GRAS). FDA has rarely had occasion to review the GRAS status of foods derived from conventionally bred plants, because these foods have been widely recognized and accepted as safe. The policy is clear, however, that in regulating foods and their byproducts derived from new plant varieties, FDA will use section 409 to require food additive petitions whenever safety questions are sufficient to warrant formal premarket review to ensure public health protection.

FDA does not generally expect that transferred genetic material itself to be subject to food additive regulation. In regulatory terms, such material is presumed to be GRAS. Substances present in food as a result of the presence of transferred genetic material, referred to as "expression products," will typically be proteins or substances produced by the action of protein enzymes, such as carbohydrates, fats, and oils. If the intended expression product differs significantly in structure, function, or composition from substances found ordinarily in food or if it has no history of safe use in food, it might not be GRAS and might require food additive regulation. Again, the 1992 policy provides guidance to producers in evaluating the safety of food that they intend to market, including criteria and analytic steps for determining whether a product is a candidate for food additive regulation and whether consultation with FDA is appropriate. Ultimately, food producers are held accountable for the safety of their products.

As of July 1999, FDA has conducted 45 final consultations under its 1992 policy, of which 16 concerned transgenic pest-protected plants (FDA 1999b). A final consultation is evidenced by a letter from FDA acknowledging completion of the consultation process. The agency likely has had many more preliminary consultations with researchers and producers during the same period, although no public record is kept of such meetings.

4.1.3   The Environmental Protection Agency and the Regulation of Pesticides

The Federal Insecticide, Fungicide, and Rodenticide Act

The Federal Insecticide, Fungicide, and Rodenticide Act (FIFRA) is a licensing statute under which EPA regulates the sale, distribution and use of pesticides (US Congress 1947). Pesticide is defined broadly as including any substance or mixture of substances intended for preventing, destroying, repelling, or mitigating a pest (US Congress 1947, section 136(u)). The concept of pesticidal intent is critical to the definition pesticides under federal law. Pest means: 1) any insect, rodent, nematode, fungus, weed, or 2) any other form of terrestrial or aquatic plant or animal life, or virus, bacterium, or other microorganism (except viruses, bacteria, or other microorganisms on or in living humans or other living animals) that the EPA declares to be a "pest" (US Congress 1947, section 136(t)).

The statute authorizes EPA to exempt a pesticide from the requirements that would ordinarily apply if the agency determines that the substance is either adequately regulated by another federal agency or of a character that is unnecessary to regulate under FIFRA to carry out the purposes of that statute (US Congress 1947, section 136 w(b)). Examples of exemptions issued by EPA are shampoo products designed to kill head lice and subject to FDA regulation as human drugs; articles treated with pesticides, such as insect-protected lumber and mildew-resistant paints, in which the pesticides are already registered for such use; and natural and synthetic pheromones when used in traps (EPA 1988a, sections 152.20b, 152.25a, and 152.25b). EPA has also issued regulations identifying substances that are not considered pesticides at all because they are not for use against pests or not used for a pesticidal effect (EPA 1988a, sections 152.8 and 152.10). Such substances include fertilizers, plant nutrients, deodorizers, and products that exclude pests by providing a physical barrier and that contain no toxicants, such as pruning paints for trees. In sharp contrast with pesticides exempted from FIFRA regulation, substances that EPA deems to fall outside the definition of a pesticide are subject to regulation under other federal statutes, such as section 409 of FFDCA (US Congress 1958) for food additives, the Toxic Substances Control Act (US Congress 1976b) for industrial and consumer chemicals, and the Consumer Product Safety Act (US Congress 1976a).

Modern genetic techniques permit the development of plants that produce their own pesticides or are otherwise resistant to insects, viruses, and other plant pests. That capability is in some respects an extension of conventional plant breeding techniques that attempt to select the heartiest and most disease-resistant strains for use in producing hybrid seeds and plants for commercial agriculture and home gardens. Plants and other macroorganisms with pesticidal properties have been exempted from the requirements of FIFRA for many years (EPA 1988a, section 152.20a). The exemption was established before any consideration of modern biotechnology to exempt the many plant species that are naturally pest-protected (such as chrysanthemums) and insects and other macroorganisms (such as lady bugs and praying mantises) that act as natural pest control agents (OSTP 1986, p. 23320). EPA refers to this entire category of products as "biological control agents."

To be registered under FIFRA, a pesticide must not cause "unreasonable adverse effects on the environment". This phrase is defined as including both ecological concerns and risks to human health. Traditionally, that criterion required EPA to balance the potential adverse effects associated with the use of compounds that are often inherently toxic against their social, economic, and environmental benefits (US Congress

1947, section 136bb(1)). Since 1996, EPA has been required to apply a safety-only standard when examining the potential dietary risks that may be posed by residues of a pesticide that might be found in food (US Congress 1947, section 136bb(2)). Registration is conditioned on the submission and review of test data regarding the health and ecological effects of the pesticidal substance.

Section 408 of the Federal Food, Drug, and Cosmetic Act

Any substance deemed to be a pesticide under FIFRA is automatically subject to regulation under FFDCA section 408 if used on a food or feed crop or if residues of it are otherwise expected to occur on food or feed (US Congress 1958). EPA's jurisdiction under FFDCA applies even if the pesticide has been exempted from regulation under FIFRA. Section 408 provides authority for EPA to issue regulations that permit pesticide residues in or on food. Maximum permissible residue levels for pesticides are referred to as tolerances and are set by rule for raw agricultural commodities and for processed food and animal feed under the same "reasonable certainty of no harm" standard that FDA applies to food additives under section 409 of FFDCA. Section 408 also authorizes EPA to issue exemptions from the requirement of a tolerance where a pesticide poses no toxicological concerns and/or dietary exposure is negligible. By definition, a pesticide cannot be a food additive.

Additional data related to dietary exposure must be submitted to EPA to support issuance of a tolerance in conjunction with the registration of a food-use pesticide. As with unapproved food additives, in the absence of a duly promulgated tolerance or exemption, or if a residue level exceeds the tolerance, the food is deemed to be adulterated and subject to enforcement action under section 402 of the FFDCA (US Congress 1958, section 342(a)). Although EPA is responsible for setting pesticide tolerances, foods are subject to inspection and enforcement action by FDA.

4.2   EVALUATION OF THE ENVIRONMENTAL PROTECTION AGENCY'S REGULATION OF PESTICIDAL SUBSTANCES IN PLANTS UNDER THE 1994 PROPOSED RULE

In 1994, after a long review of regulatory options and having gained valuable experience in the evaluation of proposals for field tests of several transgenic pest-protected plants, EPA announced its intention to regulate the pesticidal substances produced in such plants, but not the plants themselves, under the provisions of FIFRA and FFDCA (EPA 1994a, c). The committee found that

Consistent with the coordinated framework and EPA's statutory mandates, EPA has determined that pesticidal substances expressed in plants meet the statutory definition of a pesticide and has asserted jurisdiction over pesticidal substances in transgenic pest-protected plants. If such substances were not considered pesticides, they would be subject to regulation under other federal statutes.

In effect, under the 1994 proposed rule, EPA would regulate pest-protected plants in the same way that it had traditionally regulated treated articles (EPA 1988a, section 152.25a). As long as a pesticidal substance is approved, or "registered", for a given use, the treated article itself (in this case, the plant) is not subject to regulation under FIFRA. EPA's original proposal referred to these products as plant-pesticides, creating considerable confusion and controversy (Hart 1999a, b, c): some thought, and apparently still believe, that EPA was regulating the plants themselves as pesticides. The agency has recently sought public comments on the adoption of an alternative term (EPA 1999c). In summary, the committee found that

There is a misunderstanding on the part of many parties that plants themselves are being regulated by EPA as pesticides.

The committee recommends that

EPA's rule and preamble should clearly restate the agency's position that genetically modified pest-protected plants (that is, plants modified by either transgenic or conventional techniques) are not subject to regulation as pesticides. EPA must remain sensitive to the erroneous perception that plants are being regulated as pesticides.

As discussed in chapters 1 and 3, EPA's proposal included a policy statement, regulations, and a number of specific exemptions from the tolerance requirements that would ordinarily apply under FFDCA. EPA would capture pesticidal substances produced in plants by amending the long-standing FIFRA exemption for biological control agents and then exempting pesticidal substances that did not warrant review, with separate exemptions required under FFDCA. Although the proposal has not been finalized, the agency has been implementing its essential elements in registration actions taken since 1995. Field testing of plants modified to express pesticidal traits has been sanctioned by EPA case by case since as early as 1992.

EPA regulation typically proceeds in two or three distinct stages, depending on the product involved. First, researchers interested in conducting large-scale field tests (10 acres or more) apply for an experimental use permit under section 5 of FIFRA (US Congress 1947, section 136c). Generally, at this point small-scale field tests (under 10 acres) would have already been conducted pursuant to a permit or notification under USDA's plant pest program. EPA does not require permits for field tests of under 10 acres unless the crop is to be used for food or animal feed or unless the small-scale testing is not conducted pursuant to a USDA permit, notification, or deregulation determination. The next stage, which applies to most products, involves an application to EPA for a registration that is limited to the production of propagative plant products, such as seeds, tubers, corms and cuttings (EPA 1995d). The production of these plant reproductive materials is an integral step in the development of commercial plant varieties. Finally, an application for full commercialization of the plant-expressed pesticidal substance is submitted for agency review under section 3 of FIFRA. If the plant will be used for food or feed, the applicant must also petition for establishment of a tolerance or an exemption from tolerance requirements under section 408 of FFDCA. Under EPA's proposed policy, both the registration and the tolerance action apply to the pesticidal substance and the genetic material necessary for its production in the plant.

The proposed rule includes several exemptions from regulation as plant-pesticides (sections 1.5.3 and 3.2). However, it does not explicitly address the need, on the basis of new information or improved understanding of the science, to create exemptions for additional categories of pesticidal substances under FIFRA, FFDCA, or both. It also does not discuss the need to revisit existing exemptions to assess whether they should be revoked or restricted on the basis of new information or changed circumstances. The committee found that

Current law provides sufficient flexibility for agencies to regulate products on the basis of risk and/or uncertainty and to exempt from regulation products believed to pose negligible risk.

Therefore, the committee recommends that

Regulations should be considered flexible and open to change so that agencies can adapt readily to new information and improved understanding of the science that underlies regulatory decisions.

EPA should make explicit a process for the periodic review of its regulations on the basis of new information or changed circumstances to identify additional categories of pesticidal substances expressed in plants that should be exempt from regulatory requirements and existing exemptions that should be revoked or restricted.

Finally, the proposed rule would establish several exemption categories, but does not offer any opportunity for an applicant to seek an exemption for an individual product. Given the dynamic nature of the technology, products with unique characteristics and use patterns that might warrant specific exemptions probably will be developed within the next 5 to 10 years. Without a mechanism to address these individual products case by case, a time-consuming rule-making process would be required to establish one or more new exemption categories. The committee also recommends that

EPA's rule should establish a process for applicants that do not qualify for an existing exemption to consult with the agency and seek an administrative exemption on a product-by-product basis when the pesticidal substance in the plant does not warrant registration. The process should be transparent, with sufficient information made available to allow subsequent applicants to benefit.

For a substance to qualify for exemption from FIFRA requirements in the proposed rule, EPA would require any person who sells or distributes it to notify the agency of any new information concerning potential adverse effects on human health or the environment associated with the product (EPA 1994a). That provision would, for the first time, require nonregistrants to comply with a reporting obligation imposed by statute on registrants (FIFRA § 6(a)(2); US Congress 1947, section 136d(a)(2)). Although little attention has been directed to the impact of this proposal, it would probably apply to many plant breeders, researchers and seed distributors that work with conventional pest-protected plants and have never been subject to FIFRA or EPA jurisdiction. The proposed rule does not assess the potential for taking advantage of monitoring systems that use federally funded insect surveys, independent crop consultants, and USDA extension agents to identify potential adverse effects associated with conventional pest-protected plants and other crops. The committee recommends that

EPA should publicly reexamine the extent to which FIFRA adverse effects reporting is intended to apply to plant breeders, researchers, and seed distributors of conventional pest-protected plants who have never been subject to FIFRA or EPA jurisdiction. For products that meet the definition of a pesticide but are exempt from registration under FIFRA, EPA should review the extent to which existing field monitoring systems could substitute for traditional FIFRA reporting requirements.

4.3   EVALUATION OF THE REGULATION OF TRANSGENIC PEST-PROTECTED PLANTS UNDER THE MULTIAGENCY APPROACH OF THE COORDINATED FRAMEWORK

4.3.1   Overview

The US regulatory scheme for biotechnology products relies on multiple agencies to implement a mosaic of existing federal statutes. Each statute has a specific goal, for example to protect public health and the environment or to ensure food safety. The mosaic approach was deemed appropriate by the coordinated framework to regulate the diverse new biotechnology products and to provide credible assessments that would form the basis of sound regulatory determinations without unduly hindering the development of the technology.

The success of the multiagency approach can be assessed relative to three objectives:

- Sound science
- Effective coordination
- Transparency and public trust.

Scientific issues were addressed primarily in chapters 2 and 3, but their relevance to coordination, transparency, and public trust will be addressed in the discussion that follows. Only through effective coordination can the three lead agencies—EPA, USDA, and FDA—minimize duplication, avoid inconsistent regulatory decisions, address potential gaps in oversight, ensure that regulations evolve with experience and scientific advances, and effectively review the human health and environmental safety of products. Ultimately, the credibility of the regulatory process will depend heavily on the public's ability to understand the process and the key scientific principles on which it is based.

The coordinated framework addresses several elements that contribute to a sound regulatory process. The committee has considered those elements and identified five that are most relevant to the immediate task (box 4.1).

4.3.2   Coordination Under Existing Policy Statements and Proposals

The coordinated framework established several guiding principles to help the federal agencies coordinate their regulatory responsibilities. It states (OSTP 1986) that

---

Box 4.1
Elements that Support the Objectives of the
Coordinated Framework

- Consistency of definitions and regulatory scope.

- Clear establishment of lead and supporting agencies with a mechanism for effective interagency communication.

- Consistency of statements of information to support reviews.

- Comparably rigorous reviews.

- Transparency of review process.

The agencies will seek to operate their programs in an integrated and coordinated fashion and together should cover the full range of plants, animals and microorganisms derived by the new genetic engineering techniques....Agencies have agreed to have scientists from each other's staff participate in reviews.

Consistent with regulatory practice regarding traditional products, the 1986 framework called for jurisdiction over biotechnology products to be determined by their use. It identified the lead agency and supporting agencies that would be responsible for the oversight of various classes of products (table 4.1). The approach was explained as follows:

Where regulatory oversight or review for a particular product is to be performed by more than one agency, the policy establishes a lead agency, and consolidated or coordinated reviews.

Two other principles enunciated in the framework to promote coordination are that agencies should adopt, to the extent permitted by their statutory authorities, consistent definitions of the organisms subject to review; and that agencies should use reviews of comparable rigor. The authors of the policy also recognized that future scientific developments should lead to further refinements in the coordinated framework. They expected regulations to evolve as scientists and regulators gained experience in predicting which products required more or less controls.

EPA's 1994 proposed policy on pesticides subject to FIFRA and FFDCA discusses interactions with other agencies The policy makes EPA the federal agency primarily responsible for the regulation of pesticides and states that EPA works closely with USDA and FDA in fulfilling this mission. On the matter of coordination with USDA, EPA's proposed policy states (EPA 1994a, p. 60513) that

> EPA and USDA-APHIS have consulted and exchanged information on plants and plant-pesticides and intend to continue to do so in the coordination of their regulatory activities. The two Agencies also have and intend to continue to consult closely on scientific issues related to the safety considerations associated with the environmental impact of field tests of plant-pesticides.

A similar statement of commitment to coordination is made with respect to EPA-FDA interactions on jurisdictional questions and scientific matters. To minimize potential overlap, the proposed policy states that EPA will address food safety issues associated with plant-pesticides. Any food safety questions beyond those associated with plant-pesticides are under FDA's jurisdiction.

EPA has registered 10 pesticidal substances expressed in transgenic potato, cotton, or corn plants and has established corresponding exemptions from the requirement of a tolerance for these pesticidal substances under the agency's proposed regulations (see sections 1.5.3 and 3.2). Seven additional pesticidal products, also considered by EPA to be subject to its jurisdiction, are exempt from FIFRA registration because they consist of coat proteins of plant viruses. The transgenic pest-protected plants that express the exempt pesticides include potato, watermelon, zuchini, papaya, and cucumber. Indicating the shared responsibility for oversight of these products, USDA has made a determination of non-regulated status for each of the transgenic pest-protected plants. Those plants were formerly considered "regulated articles" under the FPPA.

TABLE 4.1   Regulatory Scheme for Coordinating Reviews of Commercial Biotechnology Products

Product Class	Lead Agency (Other Participating Agencies)	Federal Statutes
Plants and animals	USDA-APHIS (USDA-FSIS[a], FDA)	FPPA, PQA[b], NEPA, FFDCA
Pesticide microorganisms	EPA (USDA-APHIS)	FIFRA, FFDCA, FPPA, PQA[b], NEPA
Food and additives	FDA (USDA-FSIS[a])	FFDCA

[a]Food Saftey Inspection Service
[b]Plant Quarantine Act

The producers of the products also voluntarily engaged in consultations with FDA pertaining to the safety of the foods derived from the plants.

There are opportunities for interagency coordination during at least two stages of the regulatory process for transgenic pest-protected plant products. The first comes early in the process, when the developer is discussing the prospective product with the regulatory agencies to determine the kinds of data and information that will be necessary to support the regulatory review. These discussions are referred to as presubmission consultations and are encouraged by all three agencies. This is often the time when unique aspects of the product are discussed. A new product could raise jurisdictional questions or a need for new or different approaches to product testing or risk assessment.

Issues associated with new transgenic pest-protected plants might be of interest to more than a single agency. For example, a product consisting of a crop-gene combination that could result in gene flow and pose a potential human or environmental impact might raise legitimate issues for EPA or USDA and possibly for FDA as well. Interagency discussions at this early stage could help to avoid problems and delays later. To the committee's knowledge, the agencies have not yet interacted with one another on product-specific issues at this stage of the regulatory process.

Although such interaction would appear to benefit all parties, there could be several reasons for the apparent lack of activity. One reason might be that the product is highly confidential at this early stage of development and the producer prefers to work with each agency separately before submission. If that is the case, agencies might be unable to interact without the producer's permission because of legal constraints on the sharing of trade secrets and other confidential business information (CBI).

A second opportunity for interagency coordination is the period during formal product review, when the agencies are formulating their regulatory decisions on a product. Successful coordination during this period requires an effective infrastructure within and between agencies that promotes and rewards cooperative interaction. In being consistent with CBI requirements, all agencies attempt to provide each other with as much information as possible to facilitate communication on issues of mutual concern. EPA has taken steps to clear representatives of other agencies for access to CBI in submissions made to EPA. The ability of agencies to communicate unencumbered by CBI constraints can only enhance the credibility and public acceptance of the regulatory process.

Effective interagency coordination relies on a high degree of consistency in definitions, regulatory scope, and technical guidance of applicants, as well as effective communication and transparent review processes of comparable rigor (box 4.1). Several of those elements are highlighted in the coordinated framework (OSTP 1986), and the committee has considered each of them in its evaluation of the current status of interagency coordination in regulating transgenic pest-protected plant products. Although all the elements are desirable for promoting coordination, the committee recognizes that they might not all be relevant for every product. The committee also understands that the degree to which some of the elements are achievable is limited by the requirements of the statutes that the agencies administer. The following sections of this chapter discuss those elements outlined in box 4.1.

4.3.3   Consistency of Definitions and Regulatory Scope

To facilitate consistent and efficient regulation, the coordinated framework established the principle that agencies should adopt consistent definitions of regulated products "to the extent permitted by their respective statutory authorities." An important implication of this principle is that definitions affect the scope of products subject to regulation. Each agency must be cognizant of the scope of products delineated for regulation by its fellow agencies to ensure that regulatory coverage is coordinated and complete, but not unnecessarily duplicative. The committee found that

Although statutory constraints prevent agencies from adopting uniform definitions for certain regulatory terms, this does not appear to have unduly hindered their ability to implement meaningful regulations.

Each agency defines transgenic pest-protected plant products in terms consistent with its regulatory authority: pesticides for EPA, plant pests for USDA, and foods for FDA (table 4.2). The result is that there is no uniform interagency definition of these products. EPA focuses regulatory

TABLE 4.2   EPA, USDA, and FDA Definitions of Regulated Products and Substances

	EPA	USDA	FDA
Regulated Product	Plant-pesticide (plant-expressed protectant)	Plant pest, regulated article	Food, feed, food additive
Regulated Substance	Pesticidal substance and genetic material necessary for its production	Organism engineered to contain sequences from plant pests	Human food (whole or processed), animal feed

attention on pesticidal substances produced in plants rather than the plants themselves. These substances and the genetic material leading to their production are referred to in the 1994 proposed rule as plant-pesticides. USDA has declared some genetically engineered plants to be "regulated articles" because of potential plant pest risk. FDA regulates foods derived from new plant varieties. The lack of consistent product definitions appears to be an unavoidable outcome of regulating under existing statutes. Agencies can minimize the confusion that results from this situation by aggressively communicating how their regulations link to cover the full range of potential concerns (for example, food safety, environmental protection, and plant pest risk) for a single transgenic pest-protected plant product such as corn modified to express the *Bacillus thuringiensis* insect-control protein.

There is a more urgent need concerning consistency in the scope of transgenic pest-protected products regulated by EPA, USDA, and FDA. The scope of products covered needs to be consistent across agencies to the greatest extent possible to ensure that all products receive the appropriate oversight, and that human health and the environment are thus protected appropriately. EPA articulates a broad scope of coverage that appears to include all plant-expressed substances that meet the FIFRA definition of "pesticide," including some plantregulators (EPA 1994a). Several categories of plant-expressed pesticidal substances are then proposed to be exempt from regulation because the agency believes that they are of a type that does not require regulation under FIFRA or that they are adequately regulated by other federal agencies (sections 1.5.3 and 3.2).

FDA's regulatory coverage is similarly broad. It covers all food and feed, irrespective of how they were developed. There are no explicit exemptions from coverage, but premarket approval is not required unless a food or feed contains substances or demonstrates attributes that are not usual for the product. USDA exercises explicit regulatory authority over transgenic pest-protected plants that have been genetically engineered to contain inserted genetic material believed to have plant pest potential. All other transgenic pest-protected plants are implicitly exempt from USDA regulation unless the agency has a "reason to believe" that they could pose a plant pest risk.

Thus, all three agencies appear to have broad regulatory authority to cover transgenic pest-protected plants, but USDA and EPA have elected to narrow their effective scope of coverage by exempting particular products. The committee identified situations in which such exemptions warrant further scrutiny: the current limitation of USDA's explicit scope of oversight and EPA's proposed broad exemption of virus coat proteins under FIFRA (section 3.2.2). Both situations have the potential to result in gaps in regulatory coverage that could lead to instances where public health or environmental issues might not be adequately addressed. In general the committee found that

The scope of product reviews, as delineated by USDA and EPA, has the potential to result in gaps in regulatory coverage.

Concerning USDA's scope, USDA-APHIS oversees field tests of genetically modified crops, including transgenic pest-protected plants. It is the only agency that reviews the environmental and agricultural effects of transgenic pest-protected plants whose pesticidal substances EPA has has proposed to exempt from regulation under FIFRA. The scope of USDA's oversight includes "any organism which has been altered or produced through genetic engineering, if the donor organism, recipient organism, or vector or vector agent belongs to the genera or taxa designated in Section 340.2 and meets the definition of a plant pest" (USDA 1987, section 340.1). Many plants do not automatically meet the definition of a "plant pest." Thus, the upshot of this language is that, without a specific determination to the contrary, USDA regulations cover only genetically modified plants that have inserted genetic material from plant pests. In practice, USDA regulates genetically engineered plants with insertion vectors and promoters from plant pathogens, such as *Agrobacterium tumefaciens* and cauliflower mosaic virus. The agency also reviews voluntary submissions from those whose plants are not expressly covered.

Use of a small amount of genetic material from a plant pathogen as a vector or promoter, however, does not result in plants that pose greater plant pest risks than other types of genetically modified plants. The small amount of genetic material from plant pathogens that is inserted into plants does not result in diseased plants (Center for Science Information 1987; Goldburg 1989).

The development of new techniques for genetically engineering crops means that the scope of USDA's regulations might now fail to encompass some genetically engineered crops that the agency wishes to regulate. A number of techniques, such as the use of microprojectile guns, can now be used to insert DNA into plants without the use of the *Agrobacterium* vector. Genetic engineers can now make genetic constructs with promoters that are no longer automatically subject to USDA oversight, not because they pose any more or less plant pest risk than plants now being regulated by USDA, but simply because of the techniques used to modify them. Although companies developing such plants may voluntarily notify USDA of field tests, it remains to be seen how USDA will regulate (or deregulate) such crops when they are commercialized. Moreover, companies and researchers obviously have considerable discretion whether they continue to notify USDA of field tests without a legal requirement to do so. Therefore, the committee recommends that

USDA should clarify the scope of its coverage as there are some transgenic pest-protected plants that do not automatically meet its current definition of a plant pest

4.3.4   Clear Establishment of Lead and Supporting Agencies With a Mechanism for Effective Interagency Communication

The coordinated framework does not identify lead and supporting agencies for oversight of transgenic pest-protected plants. That is probably because research with this category of plants was relatively new when the framework was created and field testing had not yet been conducted. Instead, the coordinated framework indicates that USDA is the designated lead agency for plants and reiterates that EPA has exempted from registration, under FIFRA, plants that are biological control agents.

Although EPA's 1994 proposed policy (EPA 1994a) reiterates the exemption of plants as biological control agents, it points out that EPA will regulate pesticidal substances expressed in the plants and the genetic material necessary for the production of the substances. The policy also clearly articulates the division of jurisdiction over the substances between EPA and FDA. The policy states that EPA will address food safety issues associated with pesticidal substances, including selectable markers; FDA will be responsible for any food safety issues separate from pesticidal substances such as changes in food quality and unintended compositional changes. That clear delineation of responsibility has resulted in product reviews that avoided duplication and achieved consistency. The committee found that

The delineation of EPA and FDA jurisdiction over transgenic pest-protected plant products is generally well defined. Agency reviews generally lack duplication and achieve consistency. The agencies are working together in an effort to potentially modify jurisdiction over selectable markers in the future to reduce ambiguity and minimize the potential for duplication.

Since publication of the 1994 policy, EPA and FDA have identified selectable markers as an area where a shift in lead agency may be appropriate. Having reviewed numerous products that contain selectable markers and having received public comments on this issue, EPA published a request for comments on excluding selectable markers as pesticide inert ingredients. EPA proposed that FDA rather than EPA, have direct jurisdiction over those substances in food products. Among the reasons given for the proposed change were statutory ambiguity pertaining to EPA oversight of selectable markers and public comments asserting the potential for duplication of reviews with FDA. The committee believes that EPA's request for comments on this topic shows how regulation under the coordinated framework is continuing to evolve with experience and public input.

Although not identified as such in EPA's 1994 policy statement, responsibility for allergenicity is shared by EPA and FDA. Both agencies are responsible for addressing public health issues associated with pesticidal substances in crops that are potential food allergens. If EPA registers and establishes a tolerance for a pesticidal substance that is a potential food allergen, FDA has the authority to ensure that resulting food products carry appropriate precautionary labeling. The committee was encouraged to learn that EPA initiates consultations with FDA when issues of potential food allergenicity arise in connection with a product under review. FDA has shared with EPA its expertise on the assessment of food-allergenicity issues and has provided access to its database that is used to screen products for potential allergenic components. Therefore, the committee concludes that

EPA and FDA appropriately share responsibility for regulation of plant-expressed pesticidal substances that are potential food allergens. However, although there appears to be a high level of communication between the agencies when a potential food allergen is identified, there is no formal mechanism to ensure appropriate communication in the future as more products come under review.

Therefore, the committee recommends that

EPA and FDA develop a memorandum of understanding (MOU) that establishes a process to ensure a timely exchange of information on plant-expressed pesticidal substances that are potential food allergens. The MOU should articulate a process under which the agencies can regulate potential food allergens in a consistent fashionæby EPA through tolerance setting and by FDA through food labeling.

Neither the EPA proposed rule nor USDA's regulations provide a clear statement on the division of jurisdiction or shared responsibility between EPA and USDA for transgenic pest-protected plant products. In practice, because EPA has lead responsibility for pesticides, it has assumed the lead-agency role for those products. There is implicit recognition that EPA is the lead agency on human-health issues and most environmental issues, whereas USDA is responsible for assessing the potential for plant pest risk. The committee's discussions with EPA and USDA identified several subjects on which they request nearly identical information; in some instances, they appear to assess the same issues. That raises the question of regulatory overlap, which could lead to duplicative reviews and conflicting regulatory determinations.

The information that EPA and USDA require to support their FIFRA and FPPA risk assessments and USDA's NEPA environmental assessments are summarized in table 4.3. A comparison of EPA and USDA requirements suggests a substantial level of duplication. The committee's review of several EPA fact sheets for registered transgenic pest-protected plant products indicates that the agency requires companies to submit the results of specific laboratory studies to assess mammalian toxicology, protein digestibility, and effects on potentially exposed nontarget organisms. EPA uses this information to determine whether there is a reasonable certainty of no harm to humans consuming the plant-pesticide, as required under FFDCA; and that the product will not cause unreasonable adverse effects to human health or the environment, as required under FIFRA. For the most part, companies appear to provide summaries of these data to satisfy USDA's information needs in case of overlap. Companies might also submit the human health data on a transgenic pest-protected plant product to FDA, although FDA review is directed at the nutritional and compositional characteristics of the food and the potential for unintended alterations in food constituents.

USDA also requests applicants to provide human-health and ecological information; this suggests an unnecessary overlap in regulatory oversight. However, except for information pertaining to USDA's assessment of plant pest risk, the human health and ecological information that it receives is used to support its environmental assessment under NEPA, not to support its granting or denial of a permit or determination of nonregulated status under FPPA. USDA does not typically ask applicants to generate human health or environmental data de novo to support its NEPA findings. Instead, companies are asked to submit the available information to support the environmental assessment. Therefore, the duplication in requested information stems largely from USDA's statutory obligations under NEPA. For the most part, the duplication has allowed health and ecological issues to receive a broader assessment and has not generally led to conflicting regulatory decisions. In summary the committee concludes that

There is significant overlap in the human health and environmental information that EPA and USDA receive and evaluate in their assessment of transgenic pest-protected plant products. The duplication appears to result from NEPA requirements that apply to USDA and has not generally led to confusion or serious incidents of conflicting regulatory decisions.

TABLE 4.3   USDA and EPA Data Requirements for Assessing Effects of Transgenic Pest-Protected Plant Products

USDA[a]	EPA[b]
Information for review as regulated article	
Objective: Assess potential plant pest risk	**Objective: Assess potential for health and ecological effects**
• Genetic analysis	• Product identity (construct, characterization, markers, vectors)
• Molecular biology of transfer	• Protein digestibility
• Phenotype of article	• Mammalian toxicology (acute oral)
• Environmental consequences	• Allergenicity potential
• Description of mode of action	• Gene expression
• Current uses	• Environmental fate of protein
• Effect on weediness	• Gene transfer potential
• Gene transfer	• Nontarget organism toxicity (avian, fish, terrestrial and aquatic invertebrates)
• Potential for adverse effects	• Endangered species considerations
• Toxicology data on nontarget organisms and threatened and endangered species	
Information for environmental assessment	
Objective: Assess potential for environmental impact	
• Effect on agricultural practices	
• Potential impact of pollen escape	
• Effect on susceptibility of pathogens or insect pests	
• Effect on resistance of pests	
• Toxicology data on nontarget organisms (beneficial insects, animals, and humans)	
• Potential change in virulence (viruses)	
• Cumulative environmental effects	

[a]USDA 1996a.
[b]CFR 158.9(d); EPA (1999a,b,1998a,b, and 1999f).

However, where EPA and USDA assert regulatory authority over the same endpoint, the lack of clarity as to the lead agency and the differing bases for decision-making can, on occasion, lead to confusion both in the agencies and in the regulated community. For example, the record indicates potential confusion in instances where gene transfer is analyzed by EPA and USDA. In the case of Bt cotton USDA and EPA asked for much of the same information to assess gene-flow issues. USDA concluded that gene transfer prompted no concerns and granted deregulated status to Bt cotton without restrictions. In contrast, EPA placed geographic restrictions on the planting of Bt cotton until additional information could be provided to adequately assess the potential for and consequences of transfer of the Bt gene to related species. EPA was focusing on overall environmental impact, whereas the USDA conclusions were related to plant pest issues.

The agencies indicated that they did not communicate with one another on this issue before making their regulatory determinations. However, USDA issued its determination of nonregulated status in June 1995 and EPA registered Bt cotton four months later in October 1995. It appears that the agencies were reviewing Bt cotton during a similar period, so interagency discussions presumably could have been held. The committee recognizes that science-based decisions can depend on an agencies regulatory perspective and that decisions based on the same information can differ. For example, USDA's FPPA determinations are driven by concern about plant pest risk and crop protection, whereas EPA's FIFRA determinations hinge on the potential for adverse impacts on nontarget species and environmental protection in a general sense. In the case of Bt cotton, differentdeterminations concerning the need for geographic limits appear to have been based on somewhat different regulatory end points and levels of comfort with the available information. This may have resulted in stakeholder confusion and raised questions about the credibility of assessments.

The foregoing example emphasizes the need for agencies to avoid inadvertent duplication or the appearance of inconsistency in decisions by increasing their coordination in developing guidance in subjects of common interest and maintaining communication on data needs that are believed to be mutually exclusive. To enhance coordination , the committee recommends that

EPA, USDA, and FDA should develop a memorandum of understanding for transgenic pest- protected plant products that provides guidance to identify the regulatory issues that are the purview of each respective agency (for example, ecological risks and tolerance assessment for EPA, plant pest risks for USDA, and dietary safety of whole foods for FDA); identifies the regulatory issues for which more than one agency has responsibility (for example, gene transfer for EPA and USDA and food allergens for EPA and FDA); and establishes a process to ensure appropriate and timely exchange of information between agencies.

If differences in regulatory findings remain after agency consultations, they should be adequately explained to ensure that regulatory decisions are not in conflict and do not have the appearance of conflict. Agencies should consider using *Federal Register* notices, EPA pesticide fact sheets, press releases, and their own websites and databases to provide such explanations.

Having been commercialized only within the last 5 years, transgenic pest-protected plant products have a relatively new regulatory framework. As more and more-diverse products approach the market, new issues and issues that might be less important for conventional products might warrant attention. For example, the development of Bt transgenic plant products has brought to light issues concerning insect resistance management (section 2.9). One specific concern is the potential effect of these products on the utility of Bt foliar spray products if widespread resistance to Bt insect control proteins evolves in pest populations. Resistance management is not a new issue and is not unique to Bt crops, but it has been left largely to industry and USDA to address through research, development of best practices, educational programs for growers, and other nonregulatory mechanisms. However, EPA has taken a regulatory approach to Bt crops. It has required research and monitoring, limited geographic use of some products, imposed agricultural practices for some products, and required the development and implementation of resistance management plans that rely on high Bt dose and the establishment of refugia to minimize the onset of resistant pest populations. This new role for EPA constitutes a broad set of regulatory initiatives that will probably require substantial resources to maintain, and represents a departure from, for example, the EPA initiative under the North American Free Trade Agreement that proposes voluntary labeling for resistance management related to conventional pesticides (EPA 1999e).

In contrast with EPA's approach, USDA appears to have determined that resistance management, at least as related to Bt crops, is not a plant pest risk issue that would be appropriately addressed through regulation under FPPA. But some USDA offices are working cooperatively with EPA to establish pest management centers that would foster research, education, and nonregulatory approaches to resistance management. These pest management centers are in their infancy, and it is unclear how successful they will be. One example of an activity proposed for these centers is to develop insect resistance management strategies to pesticides expressed in transgenic pest-protected plants.

In summary, the committee found that

As more transgenic pest-protected plant products reach the market additional issues concerning their safety and effective deployment will probably come to light. Not all of them will rise to a level that warrants regulation, nor will they all be amenable to traditional regulatory solutions.

Bt crops raise an important question with regard to resistance management and the potential to affect the use of Bt foliar spray products adversely. EPA–USDA collaborative efforts to develop pest management centers offer a nonregulatory approach that could serve as a model for handling other issues that might arise in the future. EPA should continue to deal seriously with Bt resistance management and any other transgenic pest-protected plants that present similar concerns, but,

Where regulation is not warranted, agencies should look for appropriate opportunities to promote nonregulatory mechanisms to address issues associated with transgenic pest-protected plant products, including encouraging development of voluntary industry consensus standards and product stewardship programs.

4.3.5   Consistency of Statements of Information to Support Reviews

As new transgenic pest-protected plant products are developed, the kinds of information necessary to support the agencies' risk assessments and regulatory determinations continue to evolve. Although agency reviews are risk based, there are differences in data requirements and in the emphasis placed on different kinds of data. Relatively little formal, detailed guidance to applicants is available. Each agency has taken a somewhat different approach in developing and providing guidance.

EPA included in its 1994 proposed policy a section on information needs and general considerations for product development and commercialization. It provides points to consider in the development of data on product identity and characterization, human health effects, ecological effects, fate of plant-pesticides in the environment, and movement via gene flow. The committee found that

In part because EPA does not have final regulations indicating the scope of products subject to FIFRA registration, relatively little formal guidance is available to companies seeking to determine the kinds of data and information that must be developed to support EPA registration of the pesticidal substances expressed by these plants.

Nevertheless, EPA is imposing data requirements and registering products case by case, creating an urgent need for companies to know to the fullest extent possible what the requirements are. Applicants can now review the existing EPA and other guidance documents, examine what previous applicants have done, and then have a presubmission consultation to seek clarification from EPA on information needs.

FDA's guidance includes its 1992 policy statement regarding the development of foods derived from new plant varieties. That document reviews the issues to be considered in the development of a food from new plant varieties, including the consideration of issues that can prompt a need for testing or consultation with FDA. In 1997, FDA issued *Guidance on Consultation Procedures for Foods Derived from New Plant Varieties* (FDA 1997c), which summarizes nine general points the FDA recommends be addressed in the development of a safety and nutritional assessment for such products as bioengineered foods.

FDA has not, however, issued guidance on the evaluation of the potential allergenicity of proteins added to foods via genetic engineering, despite assurances that it intends to. FDA coconvened a meeting on food allergy in 1994 with EPA and USDA that brought together leaders in the field to advise the agency on evaluating the allergenicity of proteins (FDA 1994b). FDA should use the results of that meeting, other scientifically relevant reports, and later research findings to develop guidance on allergenicity. The committee recommends that

FDA should put a high priority on finalizing and releasing preliminary guidance on the assessment of potential food allergens, while cautioning that further research is needed in this area.

Publication of such guidance by FDA would be helpful both to companies consulting with FDA and to companies seeking approvals from EPA, inasmuch as EPA staff depend heavily on the expertise of FDA staff on allergenicity. For example, the committee learned of one transgenic pest-protected plant that contains an insecticidal protein that has a key biochemical characteristic of food allergens: stability in simulated gastric juices (EPA 1998c). Crops containing this protein are currently restricted to use as animal feed. Tests that the manufacturer should conduct to evaluate the potential allergenicity of this protein are not well defined, and both EPA staff and the manufacturer would benefit from guidance from FDA.

USDA has guidance documents and model submissions to help applicants determine what information is needed and how to complete a submission (USDA 1996a). The application forms provide guidance as to specific information needs, but they do not discuss the depth of information required or specifically define the methods to be used.

The committee developed a comprehensive list of data needs based on guidance documents and summaries of regulatory determinations made available by the agencies. The committee provided the agencies with a detailed consolidated list and asked them to indicate the items of most importance for their regulatory review. Individual meetings were conducted with each agency to discuss the responses.

The agency responses reveal four areas where the regulatory authori-

---

Box 4.2
Information Requirements Common to all Agencies

- *Biology of recipient:*
  —information on taxonomy, habitat, and growth characteristics.

- *Molecular biology:*
  —description of source and identity of transforming material and mode of transformation.

- *Products of inserted material:*
  —identity, characterization, purpose, and mode of action.

- *Selectable markers:*
  —identification and characterization.

---

ties have similar information needs (box 4.2): biology of recipient; molecular biology; products of inserted material; and selectable markers. These common needs might be a useful starting point for a harmonized list of data requirements. Although the agencies appear to prefer different levels of detail on these four subjects, the overall scope of information is virtually identical—an observation that the committee confirmed in meetings with EPA and USDA. Each agency needs this basic information to understand a product and conduct its assessment. The committee found that

Appropriately, EPA, USDA, and FDA request that applicants submit similar information concerning the recipient plant, molecular methods, characterization of gene products, and selectable markers.

The committee recommends that

EPA, USDA, and FDA should develop a joint guidance document for applicants that identifies the common data and information the three agencies need to characterize products (for example, biology of the recipient plant, molecular biological methods used to develop the product, identification and characterization of inserted genetic material and their product(s), and identity and characterization of selectable markers).

4.3.6   Comparably Rigorous Reviews

Agency decisions concerning transgenic pest-protected plants should be based on scientific information. The information may come from the existing scientific literature. Depending on the relevance and completeness of the existing literature, agencies may require companies to generate original data to address environmental and food safety questions. USDA and EPA do not appear to be comparably inclined to require original data to support decision-making, and therefore might not always review products with comparable rigor.

At least two published studies have analyzed the use of scientific data by USDA in making regulatory decisions about transgenic crops (Wrubel et al. 1992; Purrington and Bergelson 1995). Both studies conclude that the agency relies heavily on existing scientific literature, rather than requiring that applicants and petitioners develop new experimental data directly relevant to risks that may be posed by individual transgenic plants. Purrington and Bergelson (1995) argue that there are "serious shortcomings in the content of the petitions" approved by USDA. Another analysis (Mellon and Rissler 1995) concludes that field trials conducted under USDA's oversight produce little information of value to risk assessment when it is time to commercialize transgenic crops.

USDA's approval in 1994 of a petition to deregulate transgenic squash that contained viral coat protein genes illustrates well the agency's reliance on existing information as the basis of agency determinations. Commercialization of the squash was controversial because some believed that it would probably transfer its acquired virus-resistance genes via pollination to wild squash, which is an agricultural weed in some parts of the southern United States. An analysis commissioned by USDA strongly recommended that new data be gathered for assessment of the risks that may be posed by commercialization of the squash (Wilson 1993), but USDA largely disregarded the recommendation. The agency deregulated the squash, relying almost entirely on existing information to find that commercialization of the squash would have no significant environmental impact (section 3.1.4). As the committee recommended in chapter 3, when published data are insufficient, USDA should require original data to support agency decision-making concerning transgenic crops.

In contrast, EPA generally requires that developers of transgenic pest-protected plants provide more scientific evidence, often including new data, before it makes regulatory decisions. The squash with viral coat proteins cannot be examined for comparison, because EPA was not required to review it before it was commercialized. However, the difference between the agencies' reviews can be illustrated by examining their use of data in their decisions concerning commercialization of Bt cotton discussed above: USDA deregulated the cotton on the basis of existing information about gene flow to wild cotton, and EPA placed geographic restrictions on the planting of Bt cotton until additional data could be provided.

4.3.7 Transparency of Review Process

Background

The degree to which regulatory agencies make their regulatory processes transparent influences the acceptance of a regulatory program. Transparent regulatory processes provide a clear basis for regulatory coverage, provide clear direction to those who must comply with regulations, and assist the public in understanding how the process is intended to work. Public trust in the regulatory process is gained through transparency. When the regulatory process is applied to situations where the scientific underpinnings of the technology and its impacts are continuing to evolve, transparency is crucial to identifying how scientific knowledge is being applied in the regulatory process. One of the challenges to transparency in a regulatory process is identifying the degree to which a regulatory agency needs to protect the legitimate trade secrets of the regulated community.

From a general perspective, the coordinated framework, as implemented by the various federal agencies, has elements of transparency, but there is considerable variability among the agencies. Under the programs administered by the federal agencies implementing the framework, products of biotechnology have been commercialized in diverse sectors of the economy, and there has been reasonable public acceptance of the technology. This level of acceptance suggests public trust in the American regulatory system generally and other factors such as confidence in American agriculture to produce a safe food supply. That confidence contrasts with the skepticism concerning genetic engineering in general in Europe and other parts of the world (Layman 1999; Prakash 1999). Where public trust in the current framework appears to be fragile, lack of transparency in the process can be an exacerbating factor.

The strengths and weaknesses of the framework for regulation of transgenic pest-protected plant products can be examined in the context of a transparent regulatory process. The following analysis examines the transparency of the regulatory approaches taken by the three lead federal agencies from the standpoint of the regulated community, the state-level coregulator, and the public at large. The committee found that in general,

Ready access to information on product reviews and approvals and a meaningful opportunity for stakeholder participation are critical to the credibility of the regulatory process.

Transparency at the Animal and Plant Health Inspection Service

USDA has successfully used electronically accessible databases to improve the transparency of its regulatory process and to keep the public and the regulated community informed about changes in regulation. The APHIS Biotechnology Index, on the USDA website (USDA 1999b), provides timely access to a number of databases and other information that assist researchers, companies, and the public in working with and understanding the USDA regulatory program. For example, the Biotechnology Permits Database (USDA 1999c) is updated daily and provides detail on and the current status of recent applications for movement permits, notifications of intended release, and release permits. Other accessible databases linked to the Biotechnology Index include historical environmental releases back to 1987, decision documents (environmental assessments and determinations on nonregulated status), public notices (proposed rules and links to the *Federal Register*), and summaries of field releases by type of crop, phenotype, and location. Other website resources listed in this index include guidance on applying for permits, making notifications, and petitioning for determination of nonregulated status and a variety of biosafety resource materials. The more traditional method of *Federal Register* notices to present regulations and convey regulatory decisions is also used. Those around the world who are interested in agricultural biotechnology use the databases maintained by USDA to track applications. Resource limitations, such as funding, can hamper the agency's ability to maintain the databases on those various aspects of the regulatory process. The committee finds the

USDA database on FPPA decisions to be particularly useful and user-friendly. It should serve as a model for the other agencies; the committee recommends full funding for the maintenance of the existing USDA databases.

USDA has identified aspects of data submissions that applicants may declare as CBI. In the preamble to the initial regulations, the agency directed that applicants provide a detailed statement regarding why submitted information should be treated as confidential because of the competitive harm that might result from disclosure (USDA 1987). The agency requests two copies of applications and notifications, one with CBI deleted so that the document can be shared with state coregulators. State regulators have the opportunity to assess the plant-pest risk issues for their state for permits, notifications, and determinations for deregulated status and provide comment to USDA. In response to states' concerns that applicants were designating most submitted information CBI, the agency has provided clarification on kinds of submissions that may not be so designated (USDA 1999c).

Transparency at the Environmental Protection Agency

Before EPA's Office of Pesticide Programs (OPP) issued its proposed rule in November 1994 (EPA 1994a), the agency discussed its regulatory direction with the Scientific Advisory Panel (SAP), an external scientific advisory body for OPP on matters related to FIFRA and pesticide tolerance issues under FFDCA (SAP 1994). The proposal included a policy statement that generally laid out the basis for the rule and aspects of EPA's regulatory approach to this wholly new kind of pesticide regulation. The agency began providing regulatory coverage to some plant-pesticides before the publication of the proposed rule (EPA 1994a, b). The availability of information on the regulatory program is discussed below from the standpoint of the interested public and the affected regulatory community.

Beyond the traditional means of communicating its regulatory decisions on new plant-pesticides through the *Federal Register*, EPA has used its website to improve the transparency of its regulatory program. The website provides access to pesticide fact sheets, which summarize the kinds of data and risk issues evaluated by the agency for individual active ingredients in making regulatory determinations, and it links to *Federal Register* notices of regulatory determinations under FIFRA and FFDCA (EPA 1999b). More-detailed evaluations of submitted data are not available on the website but can be requested under the Freedom of Information Act. EPA is not always able to respond to such requests in a timely manner, however, and the committee recommends that

EPA make data evaluations readily available on its website or in response to written requests.

The website provides text of presentations by EPA officials, which contain details of current regulatory approaches to plant-pesticides. (EPA 1999b). This latter resource, along with EPA fact sheets, is currently the best source of information on the kinds of data that the agency is asking for to address the array of substantive risk issues posed by plant-pesticides. EPA has also posted information and papers regarding issues on resistance management related to Bt crops (EPA 1998d and 1999b); this information is an extension of discussions between EPA and the SAP.

With respect to public communication of the health-testing results, the committee found that

The EPA pesticide fact sheets are the most readily available sources of information on human health effects, but they are not transparent with respect to either the tests performed or the results of the tests.

As these documents may be drafted to be accessible to nonexperts, sometimes they give the impression that the studies were not rigorous. For example, the fact sheet on the Bt *tolworthi* protein expressed in corn (EPA 1998c) describes toxic endpoints in one male and eight female mice in the acute-toxicity test and then states "CLASSIFICATION: Acceptable." The basis for that classification with some details of the design of the tests, the number of animals involved, and other testing methods should have been presented so that the public could appropriately evaluate the scientific rigor of the test. Another fact sheet, on Bt Cry3A in potato (EPA 1995a), fails to state the number of animals tested; however, details provided by the registrant (Lavrik et al. 1993) make it clear that the tests, although minimal, included an adequate number of animals. Synopses of the methods and data from which the information is obtained would be valuable to the readers. Therefore, the committee recommends that

EPA pesticide fact sheets should be prepared with greater clarity and with more factual information to clearly and quantitatively present the results of safety testing.

EPA addressed the issue of CBI in its proposed rule, and proposes to require substantiation at the time a claim is made (EPA 1994a). In the proposed rule, EPA actively admonished applicants to minimize the amount of data and other information claimed as CBI. Because of inherent differences in their regulatory systems, EPA does not share applications for pesticides with state coregulators as does USDA, so a comparison of treatments of CBI claims is not possible from that perspective. However, EPA does discuss some risk issues related to plant-pesticides with the SAP in public fora (for example, SAP 1994); through that venue, it is possible to assess that the agency has not allowed broad CBI claims. EPA staff report that some registrants' attempts to make broad CBI claims have been rebuffed by the agency (EPA 1999g).

Because EPA's proposed rule is not yet final, the agency has not provided specific guidance to the regulated community on the various aspects of the regulatory approach (Andersen and Milewski 1999). The regulated community under the proposed rule includes academic researchers, plant breeders, and seed companies and is substantially more diverse than registrants of traditional pesticides. The agency has endeavored to communicate with the broader group through presentations at national meetings and has tried to work closely with groups or individuals seeking clarification of proposed exemptions and guidance on making an application (Milewski 1997; Andersen and Milewski 1999). Registrants of traditional pesticides that have expanded their scope of business to include transgenic pest-protected plant products are better prepared to respond to the new regulatory coverage because of their familiarity with the existing system. More specificity on the regulatory process is available through individual staff identified on the website. The division managing the registration of plant-pesticides would benefit from having an ombudsperson to advise potential registrants, modeled after similar positions in other OPP divisions that register chemical pesticides.

The absence of clear guidance beyond the proposed rule itself on the following three subjects detracts from the transparency of EPA's regulatory programs: how to determine more definitively whether a plant-pesticide qualifies for the proposed exemptions, how to seek exemptions under FIFRA or FFDCA, and what specific kinds of data or rationale are needed by the agency to execute its regulatory program. This lack of transparency affects not only potential registrants or others affected by the proposed rule, but also affects state pesticide co-regulators and the public in understanding how the regulatory coverage is intended to work. Generally, it appears that OPP is handicapped in its efforts to make a transparent regulatory process by lack of a final rule on plant-pesticides.

The committee recommends that

EPA should promptly complete the process for issuing regulations, policies, and guidance that set out the system of review and regulatory parameters for pesticidal substances in transgenic pest-protected plants.

Clarity is critical in these issuances, and the agency should avoid the tendency to automatically fall back on policies and procedures that apply to traditional chemical pesticides. For example, EPA should move quickly to issue guidance on the data required for pesticidal substances in transgenic pest-protected plants regulated under FIFRA and FFDCA.

Transparency at the Food and Drug Administration

Under the coordinated framework, FDA considers some aspects of transgenic pest-protected plants under the general food safety clause and other provisions of FFDCA (section 4.1.2). With the exception of determining that it may require labeling for an allergenic plant-pesticide, FDA defers to EPA for evaluation of the pesticide component of transgenic pest-protected plants. FDA established guidance under the coordinated framework when it published its policy on novel foods in 1992 (FDA 1992). The policy provided direction to the regulated community and the public about when there was a need for consultation in lieu of submission of a food additive petition.

FDA has used its website to provide direction on how to use the policy to determine when a consultation should be used, what the expectations of the agency are for safety determinations, and how the consultation process works (FDA 1999a). The website also contains a list of completed consultations (FDA 1999b), which states the products and companies involved in the consultation.

However, the details of these consultations are not readily available for public scrutiny. If the public wants to obtain documents containing information and data submitted to FDA for consultation, they must request the documents from FDA through the Freedom of Information Act (FOIA). Processing and fulfillment of FOIA requests can often take a long time.

FDA maintains an internal database on the amino acid sequence of known human allergens that has been useful to both EPA and FDA in evaluating the potential allergenicity of plant-expressed pesticides and food additives. The database is not publicly accessible, thus making it more difficult for researchers and developers to assess allergenicity. FDA and EPA generally discuss how such information is used to assess allergenicity potential in the summaries of their evaluations (FDA 1999b). However, funding constraints might affect FDA's ability to maintain and update this database as new information becomes available.

FDA does not directly address the issue of substantiation of CBI claims for novel foods derived from biotechnology. Like EPA, FDA does not have an explicit relationship with state regulators in this arena (as USDA-APHIS has with its state counterparts), so a perspective on its screening of CBI claims is not possible.

Integration of Information

The Internet has greatly enhanced agencies' ability to communicate their regulatory process to the regulated community and the public. The federal agencies involved in regulating transgenic pest-protected plant products have used this medium to varied degrees, as indicated above. However, although agencies provide cross-links to one another's resources (USDA 1999e), there is no current way to link the decisions that various agencies have made about individual plant products under their own statutes.

To improve transparency, the committee recommends that

To fulfill the intent of the coordinated framework, a database to link agencies' decisions should be developed to benefit a wide array of interested parties that are following developments in agricultural biotechnology. Such a database would enhance the information now provided by the agencies and the overall credibility of the framework. Alternative or varied funding mechanisms should be explored to maintain this database.

The above database should expand on the existing USDA-sponsored coordinated framework database (USDA 1999e) to include more public information about specific products and to link agencies' decisions about specific products.

With respect to CBI and public access to information, the committee found that

Consistent with protections afforded by law to trade secrets and CBI, agencies have made a considerable amount of information on product reviews and approvals available but there is room for improvement.

The committee recommends that

EPA, USDA, and FDA should require substantiation of CBI claims at the time of data submission.

4.4   IMPACTS OF THE COORDINATED FRAMEWORK

The impacts of the coordinated framework are likely diverse and difficult to characterize and quantify. Potential benefits associated with the regulation of transgenic pest-protected plants include increased health and environmental safety and consumer confidence. Direct costs of regulation include expenditures on additional testing (that is, above and beyond testing that would occur in the absence of regulation) and employee time spent overseeing the regulatory process and interacting with agencies' staff. They also include costs associated with delays in development and commercialization of products. If those direct costs are sufficiently high, they can increase the potential size of the market (expected sales) needed to break even and thus justify investment in a new plant variety. As a result, some crop varieties (in particular minor crop varieties) may not be developed.

The committee reviewed an analysis of the costs associated with the regulation of pest-protected plants which was authored by one of its members (see appendix A[4]).   From this analysis, the committee found that regulation of transgenic pest-protected plants under the coordinated framework and EPA's proposed plant-pesticide rule might affect small to medium-size seed companies, public sector breeders, and other small developers who are not accustomed to the testing and regulatory submissions.   Therefore, the committee recommends that

Regulators should be sensitive to the unique issues facing researchers, plant breeders, and seed distributors, particularly those in the public sector or those who have not traditionally been subject to federal regulation.

Regulatory agencies should aggressively seek to reduce regulatory costs for small biotechnology startup companies, small to medium size seed companies, and public sector breeders by providing flexibility with respect to data requirements, considering fee waivers wherever possible, and helping these parties navigate their regulatory system.

4.5   RECOMMENDATIONS

• EPA's rule and preamble should clearly restate the agency's position that genetically modified pest-protected plants (that is, plants modified by either transgenic or conventional techniques) are not subject to regulation as pesticides. EPA must remain sensitive to the erroneous perception that plants are being regulated as pesticides.

• EPA should make explicit a process for the periodic review of its regulations on the basis of new information or changed circumstances to identify additional categories of pesticidal substances expressed in plants that should be exempt from regulatory requirements and existing exemptions that should be revoked or restricted.

• EPA's rule should establish a process for applicants that do not qualify for an existing exemption to consult with the agency and seek an administrative exemption on a product-by-product basis when the pesticidal substance in the plant does not warrant registration.   The

---

[4]This appendix was authored by an individual committee member and is not part of the committee's consensus report.   The committee as a whole may not necessarily agree with all of the contents of appendix A.

process should be transparent, with sufficient information made available to allow subsequent applicants to benefit.

• EPA should publicly reexamine the extent to which FIFRA adverse effects reporting is intended to apply to plant breeders, researchers, and seed distributors of conventional pest-protected plants who have never been subject to FIFRA or EPA jurisdiction. For products that meet the definition of a pesticide but are exempt from registration under FIFRA, EPA should review the extent to which existing field monitoring systems could substitute for traditional FIFRA reporting requirements.

• USDA should clarify the scope of its coverage as there are some transgenic pest-protected plants that do not automatically meet its current definition of a plant pest.

• EPA and FDA should develop a memorandum of understanding (MOU) that establishes a process to ensure a timely exchange of information on plant-expressed pesticidal substances that are potential food allergens. The MOU should articulate a process under which the agencies can regulate potential food allergens in a consistent fashion— by EPA through tolerance setting and by FDA through food labeling.

• EPA, USDA, and FDA should develop a memorandum of understanding for transgenic pest-protected plant products that provides guidance to identify the regulatory issues that are the purview of each respective agency (for example, ecological risks and tolerance assessment for EPA, plant pest risks for USDA, and dietary safety of whole foods for FDA); identifies the regulatory issues for which more than one agency has responsibility (for example, gene transfer for EPA and USDA and food allergens for EPA and FDA); and establishes a process to ensure appropriate and timely exchange of information between agencies.

• Where regulation is not warranted, agencies should look for appropriate opportunities to promote nonregulatory mechanisms to address issues associated with transgenic pest-protected plant products, including encouraging development of voluntary industry consensus standards and product stewardship programs.

• FDA should put a high priority on finalizing and releasing preliminary guidance on the assessment of potential food allergens, while cautioning that further research is needed in this area.

- EPA, USDA, and FDA should develop a joint guidance document for applicants that identifies the common data and information the three agencies need to characterize products (for example, biology of the recipient plant, molecular biological methods used to develop the product, identification and characterization of inserted genetic material and their product(s), and identity and characterization of selectable markers).

- The USDA database on FPPA decisions is particularly useful and user-friendly, and should serve as a model for the other agencies. The committee recommends full funding for the maintenance of existing USDA databases.

- EPA should make data evaluations readily available on its website or in response to written requests.

- EPA pesticide fact sheets should be prepared with greater clarity and with more factual information to clearly and quantitatively present the results of safety testing.

- EPA should promptly complete the process for issuing regulations, policies and guidance that set out the review and regulatory parameters for pesticidal substances in transgenic pest-protected plants.

- To fulfill the intent of the coordinated framework, a database to link agencies' decisions for particular products would benefit a wide array of interested parties that are following developments of agricultural biotechnology. Such a database would enhance the existing information provided by the agencies and the overall credibility of the framework. Alternative funding mechanisms should be explored to maintain this database.

- EPA, USDA, and FDA should require substantiation of CBI claims at the time of data submission.

- Regulatory agencies should aggressively seek to reduce regulatory costs for small biotechnology startup companies, small to medium size seed companies, and public sector breeders by providing flexibility with respect to data requirements, considering fee waivers wherever possible, and helping these parties navigate their regulatory system.

References

Abdalla, O. A., P. R. Desjardins, and J. A. Dodds. 1991. Identification, disease incidence and distribution of viruses infecting peppers in California. Plant Dis. 75:1019-1023.

Akeley, R. V., W. R. Mills, C. E. Cunningham, and J. Watts. 1968. Lenape: a new potato variety high in solids and chipping quality. American Potato Journal 45:142-145.

Al-Kaff, N. S., S. N. Covey, M. M. Kreike, A. M. Page, R. Pinder, and P. J. Dale. 1998. Transcriptional and posttranscriptional plant gene silencing in response to a pathogen. Science 279:2113-2115.

Ames, B. N., M. Profet, and L. S. Gold. 1990a. Dietary pesticides (99.99% all natural). Proc. Natl. Acad. Sci. USA 87:7777-7781.

Ames, B. N., M. Profet, and L. S. Gold. 1990b. Nature's chemicals and synthetic chemicals: Comparative toxicology. Proc. Natl. Acad. Sci. USA 87:7782-7786.

Andersen, J., and E. Milewski. 1999. Regulation of Plant-Pesticides: Current Status. Presented at the International Conference on Emerging Technologies in IPM, March 8, 1999, Raleigh, North Carolina.

Andow, D., and W. Hutchinson. 1998. Bt-corn resistance management. Pp. 19-66 in Now or Never: Serious New Plans to Save a Natural Pest Control, M. Mellon and J. Rissler, eds. Cambridge, MA: Union of Concerned Scientists.

Araji, A. A., F. C. White, and J. F. Guenther. 1995. Spillovers and the returns to agricultural research for potatoes. J. Agric. Resour. Econ. 20:263-276.

Arias, D. M., and L. H. Rieseberg. 1994. Gene flow between cultivated and wild sunflower. Theor. Appl. Genet. 89:655-660.

Arnold, M. L. 1997. Natural Hybridization and Evolution. New York: Oxford University Press.

Arriola, P. E., and N. C. Ellstrand. 1996. Crop-to-weed gene flow in the genus Sorghum (Poaceae): spontaneous interspecific hybridization between johnsongrass, Sorghum halepense, and crop sorghum, S. bicolor. Am. J. Bot. 83:1153-1160.

Arriola, P. E. and N. C. Ellstrand. 1997. Fitness of interspecific hybrids in the genus Sorghum: persistence of crop genes in wild populations. Ecol. Appl. 7:512-518.

Arumuganathan K., and E. D. Earle. 1991. Nuclear DNA content of some important plant species. Plant Mol. Biol. Rep. 9:208-219.

Babcock, B. A., and W. E. Foster. 1991. Measuring the potential contribution of plant breeding to crop yields: Flue-cured tobacco 1954-1987. Am. J. Agric. Econ. 73:850-859.

Baenziger, P. S., and C. J. Peterson. 1992. Genetic variation: its origin and use for breeding self-pollinated species. Pp. 69-92 in Plant Breeding in the 1990s, March 1991 Symposium Proceedings, T. M. Stalker, and J. P. Murphy, eds. Wallingford, United Kingdom: C.A.B. International.

Baker, B., P. Zambryski, B. Staskawicz, and S. P. Dinesh. 1997. Signaling in plant-microbe interactions. Science 276:726-733.

Ball, V. E., J. C. Bureau, R. Nehring, and A. Somwaru. 1994. Agricultural productivity revisited. Am. J. Agric. Econ. 79:1045-1063.

Barham, W. S. 1953. The inheritance of a bitter principle in cucumbers. Proc. Amer. Soc. Hort. Sci. 62:441-442.

Barrett, S. C. H. 1983. Crop mimicry in weeds. Econ. Bot. 37:255-282.

Bartnicki, D. E., P. B. Lavrik, R. M. Leimgruber, C. E. Smith, and S. R. Sims. 1993a. Equivalence of microbially-produced and plant-produced *B.t.t.* protein also called Colorado potato beetle active protein from *Bacillus thuringiensis* subsp. *tenebrionis*. Submitted to EPA for Monsanto Company's registration for NatureMark NewLeaf potato.

Bartnicki, D. E., R. M. Leimgruber, P. B. Lavrik, C. E. Smith, and S. R. Sims. 1993b. Characterization of the major tryptic fragment from Colorado potato beetle active protein from *Bacillus thuringiensis* subsp. *tenebrionis* (*B.t.t.*). Submitted to EPA for Monsanto Company's registration for NatureMark NewLeaf potato.

Bartsch, D., M. Schmidt, M. Pohl-Orf, C. Haag, and I. Schuphan. 1996. Competitiveness of transgenic sugar beet resistant to beet necrotic yellow vein virus and potential impact on wild beet populations. Mol. Ecol. 5:199-205.

Bateman, A. J. 1947a. Contamination of seed crops. I. Insect pollination. J. Genet. 1:257.

Bateman, A. J. 1947b. Contamination of seed crops. II. Wind pollination. Heredity 1:235-246.

Bateman, A. J. 1947c. Contamination of seed crops. III. Isolation by distance. Heredity 1:303-335.

Baulcombe, D. C. 1996. Mechanisms of pathogen derived resistance to viruses in transgenic plants. Plant Cell 8:1833-1844.

Beachy, R. N. 1997. Mechanisms and applications of pathogen derived resistance in transgenic plants. Curr. Opin. Biotech. 8:215-220.

Beachy, R. N., and M. Bendahmane. 1998. Pathogen derived resistance and reducing the potential to select viruses with increased virulence. Pp. 87-94 in Gene Escape and Pest Resistance, R. W. F. Hardy, and J. B. Segelken, eds. Report No. 10. Ithaca, NY: National Agricultural Biotechnology Council.

Bendahmane, M., J. H. Fitchen, G. M. Zhang, and R. N. Beachy. 1997. Studies of coat protein-mediated resistance to tobacco mosaic tobamovirus: Correlation between assembly of mutant coat proteins and resistance. J. Virol. 71:7942-7950.

Berg, P., D. Baltimore, and H. W. Boyer. 1974. Potential biohazards of recombinant DNA molecules. Science 185:303.

Berg, P., D. Baltimore, and S. Brenner. 1975. Asilomar conference on recombinant DNA molecules. Science 188:991-994.

Bergelson, J., C. B. Purrington, and G. Wichmann. 1998. Promiscuity in transgenic plants. Nature 395:25.

Bergelson, J., J. Winterer, and C. B. Purrington. In press. Ecological impacts of transgenic crops. In Biotechnology and Genetic Engineering of Plants, V. Malik, ed. New York: Oxford University Press.

Bergman, J. M., and W. M. Tingey. 1979. Aspects of interaction between plant genotypes and biological control. Bull. Ent. Soc. Am. 25:275-279.

184 GENETICALLY MODIFIED PEST-PROTECTED PLANTS: SCIENCE AND REGULATION

Bernstein, I. L., J. A. Bernstein, M. Miller, S. Tierzieva, D. I. Bernstein, Z. Lummus, M. K. Selgrade, D. L.Doerfler, and V. L. Seligy. 1999. Immune responses in farm workers after exposure to *Bacillus thuringiensis* pesticides. Envir. Health Persp. 107:575-582.

Boethel, D. J., and R. D. Eikenbary (eds.). 1986. Interactions of plant resistance and parasitoids and predators of insects. Chichester, West Sussex: Ellis Harwood Press.

Borja, M., T. Rubio, H. B. Scholthof, and A. O. Jackson. 1999. Restoration of wild-type virus by double recombination of tombusvirus mutants with a host transgene. Mol. Plant-Microbe Interact. 12:153-162.

Bottrell, D. G., and P. Barbosa. 1998. Manipulating natural enemies by plant variety selection and modification: A realistic strategy? Annu. Rev. Entomol. 43:347-367.

Boudry, P., M. Morchen, P. Sanmitou-Laprade, P. Vernet, and H. Van Dijk. 1993. The origin and evolution of weed beets: Consequences for the breeding and release of herbicide-resistant transgenic sugar beets. Theor. Appl. Genet. 87:471-478.

Brennan M. F., C. E. Pray, and A. Courtmanche. 1999. Impact of Industry Concentration on Innovation in the U.S. Plant Biotech Industry. Rutgers University: Department of Agricultural, Food, and Resource Economics.

Bridges, D. C., ed. 1992. Crop Losses Due to Weeds in the United States. Champaign, IL: Weed Science Society of America.

Brower, L. P., J. N. Seiber, C. J. Nelson, S. P. Lynch, and M. M. Holland. 1984. Plant determined variation in the cardenolide content, thin-layer chromatography profiles, and emetic potency of monarch butterflies, *Danaus plexippus*, reared on the milkweed plants in California: Two *Asclepias speciosa*. J. Chem. Ecol. 10:601-639.

Brown, A. H. D., S. C. H. Barrett, and G. F. Moran. 1985. Mating system estimation in forest trees: Models, methods, and meanings. Pp. 32-49 in Population Genetics in Forestry, H. R. Gregorius, ed. New York: Springer-Verlag.

Bull, L. B., C. C. J. Culvenor, and A. T. Dick. 1968. The Pyrrolizidine Alkaloids. New York: Wiley & Sons.

Burdon, J. J. 1987. Diseases and Plant Population Biology. New York: Cambridge University Press.

Burdon, J. J., D. C. Abbott, A. H. D. Brown, and J. S. Brown. 1994. Genetic-structure of the scald pathogen (*Rhynchosporium secalis*) in Southeast Australia—implications for control strategies. Aust. J. Agric. Res. 45:1445-1454.

Burdon, J. J., and A. M. Jarosz. 1988. The ecological genetics of plant-pathogen interactions in natural communities. Philos. T. Roy. Soc. B 321:349-363.

Burks, A. W., and -R. L. Fuchs. 1995. Assessment of the endogenous allergens in glyphosate-tolerant and commercial soybean varieties. J. Allergy Clin. Immunol.1995 Dec; 96(6 Pt 1):1008-10).

Burnside, O. 1968. Control of wild cane in soybean. Weed Sci. 16:18-22.

Buschges, R., K. Hollricher, R. Panstruga, G. Simons, M. Wolter, A. Frijters, R. vanDaelen, T. vanderLee, P. Diergaarde, J. Groenendijk, S. Topsch, P. Vos, F. Salamini, and P. Schulze-Lefert. 1997. The barley Mlo gene: A novel control element of plant pathogen resistance. Cell 88:695-705.

Cai, D., M. Kleine, S. Kifle, H. J. Harloff, N. N. Sandal, K. A. Marcker, R. M. Klein-Lankhorst, E. M. J. Salentijn, W. Lange, W. J. Stiekema, U. Wyss, F. M. W. Grundler, and C. Jung. 1997. Positional cloning of a gene for nematode resistance in sugar beet. Science 275:832-834.

Carozzi, N., and M. Koziel, eds. 1997. Advances in Insect Control: The Role of Transgenic Plants. London: Taylor & Francis.

Carrasco-Tauber C., and L. J. Moffitt. 1992. Damage control econometrics: functional specification and pesticide productivity. Amer. J. Ag. Econ. 74:158-162.

Carrington, J. C., and S. A. Whitham. 1998. Viral invasion and host defense: Strategies and counter strategies. Curr. Opin. Plant Biol. 1:336-341.

CAST (Council for Agricultural Science and Technology). 1998. The proposed EPA plant pesticide rule. Issue Paper No. 10, October. [Online]. Available: http://www.cast-science.org/epar_ip.htm [1999, July 8].

Center for Science Information. 1987. Regulatory Considerations: Genetically-Engineered Plants. Summary of a workshop at the Boyce-Thompson Institute for Plant Research at Cornell University.

Chambers R. G. ,and E. Lichtenberg. 1994. Simple econometrics of pesticide productivity. Amer. J. Ag. Econ. 76:407-418.

Chambers R. G., and E. Lichtenberg. 1995. Estimating Aggregate Pest Damage in U.S. Agriculture. Working Paper No. 93-04, Department of Agricultural and Resource Economics. College Park: University of Maryland.

Chesapeake Bay Program. 1995. The State of the Chesapeake Bay, 1995. Annapolis, MD: US Environmental Protection Agency, Chesapeake Bay Program.

Chèvre, A. M., F. Eber, A. Baranger, M. Renard. 1997. Gene flow from transgenic crops. Nature 389:924.

Chèvre, A. M., F. Eber, and M. Renard. 1999. Gene flow from oilseed rape to weeds. Pp. 125-130 in Gene Flow and Agriculture: Relevance for Transgenic Crops; Proceedings of a Symposium Held at the University of Keele, Staffordshire 12-14 April 1999/ Chaired by P. J. W. Lutman. British Crop Protection Council, symposium proceedings No. 72. Farnham: BCPC.

Cohen, S., A. Chang, H. Boyer, and R. Helling. 1973. Construction of biologically functional bacterial plasmids *in vitro*. Proc. Natl. Acad. Sci. USA 70:3240-3244.

Colwell, R. K., E. A. Norse, D. Pimental, F. E. Sharples, and D. Simberloff. 1985. Genetic engineering in agriculture. Science 229:111-112.

Comai, L., D. Facciotii, W. R. Hiatt, G. Thompson, R. E. Rose, and D. M. Stalker. 1985. Expression in plants of a mutant aroA gene from *Salmonella typhirium* confers tolerance to glyphosate. Nature 317:741-744.

Cook, R. J. 1999. Experience with agricultural biodiversity: baseline for guidance in the liability and labeling debates. Paper presented at a National Research Council's Committee on Genetically Modified Pest protected Plants Public Workshop, May 24.

Council of Economic Advisers. 1999. Economic Report of the President. Washington, D.C.: US Government Printing Office.

Cowling K., and M. Waterson. 1976. Price-cost margins and market structure. Economica 43:267-274.

Cox, T. S., and J. H. Hatchett. 1986. Genetic model for wheat/Hessian fly (Diptera: Cecidomyiidae) interaction: Strategies for deployment of resistance genes in wheat cultivars. Environ. Entomol. 15:24-31.

Cox, T. S., R. G. Sears, R. K. Bequette, and T. J. Martin. 1993. Germplasm enhancement in winter wheat ¥ *Triticum tauschii* backcross poulations. Crop Science 35:913-919.

Cox, T. S., W. J. Raupp, and B. S. Gill. 1994. Leaf rust-resistance genes Lr41, Lr42, and Lr43 transferred from *Triticum tauschii* to common wheat. Crop Sci. 34:339-343.

Cox, T. S., R. K. Berquette, R. L. Bowden, and R. G. Sears. 1997. Grain yield and breadmaking quality of wheat lines with the leaf rust resistance gene Lr41. Crop Sci. 37:154-161.

Crawley, M. J., R. S. Hails, M. Rees, D. Kohn, and J. Buxton. 1993. Ecology of transgenic oilseed rape in natural habitats. Nature 363:620-623.

Crecchio, C., and G. Stotzky. 1998. Insecticidal activity and biodegradation of the toxin from *Bacillus thuringiensis* sp. *kurstaki* bound to humic acids from soil. Soil Biol. Biochem. 30:463-470.

Cresswell, J. E., A. P. Bassom, S. A. Bell, S. J. Collins, and T. B. Kelly. 1995. Predicted pollen dispersal by honey-bees and three species of bumble bee foraging on oilseed rape: A comparison of three models. Funct. Ecol. 9:829-842.

DaCosta, C. P., and C. M. Jones. 1971. Cucumber beetle resistance and mite susceptibility controlled by the bitter gene in *Cucumis sativus* L. Science 172:1145-1146.

Daniell, H., R. Datta, S. Varma, S. Gray, and S. B. Lee. 1998. Containment of herbicide resistance through genetic engineering of the chloroplast genome. Nat. Biotechnol. 16:345-348.

Decker, D. S. 1988. Origin(s), evolution, and systematics of *Cucurbita pepo* (Cucurbitaceae). Econ. Bot. 42:4-15.

Dewey, R. E., J. N. Siedew, D. H. Timothy, and C. S. Levings III. 1988. A 13-kilodalton maize mitochondrial protein in *E. Coli* confers sensitivity to *Biopolaris maydis* toxin. Science 239:293-295.

Diawara, M. M., and J. T. Trumble. 1997. Linear furanocoumarins. Pp. 175-188 in Handbook of Plant and Fungal Toxicants, J. P. D'Mello, ed. Boca Raton, Florida: CRC Press.

Dicke, M. 1996. Crop plants affect biological control agents: Prospects for breeding for host plant resistance. Pp. 72-80 In Breeding for Resistance to Insects and Mites, P. R. Ellis, and J. Freuler, eds. International Organisation for Biological Control / West Palaearctic Regional Section (IOBC/WPRS) Bulletin 19(5).

Dickson, J. O., and R. R. King. 1978. The transfer of pyrrolizidine alkaloids from *Senecio jacobaea* into the milk of lactating cows and goats. Pp. 201-209 in Effects of Poisonous Plants on Livestock, R. F. Keeler, K. R. Van Kampen, and L. F. James, eds. New York: Academic Press.

Dirzo R., and J. L Harper. 1982. Experimental studies on slug-plant interactions. IV. The performance of cyanogenic and acyanogenic morphs of *Trifolium repens* in the field. J. Ecol. 70 119-138.

Donegan K. K., C. J. Palm, V. J. Fieland, L. A. Porteous, L. M. Ganio, D. L. Schaller, L. Q. Bucao, and R. J. Seidler. 1995. Changes in levels, species, and DNA fingerprints of soil microorganisms associated with cotton expressing the *Bacillus thuringiensis* var. *kurstaki* endotoxin. Appl. Soil Ecol. 2:111-124.

Donegan K. K., D. L. Schaller, J. K. Stone, L. M. Ganio, G. Reed, P. B. Hamm, and R. J. Seidler. 1996. Microbial populations, fungal species diversity and plant pathogen levels in field plots of potato plants expressing the *Bacillus thuringiensis* var. *tenebrionis* endotoxin. Transgenic Res. 5:25-35.

Duffey, S. S. 1980. Sequestration of plant natural products by insects. Annu. Rev. Entomol. 25:447-477.

Duvick D. N. 1984. Genetic contributions to yield gains of U.S. hybrid maize, 1930 to 1980. Pp. 15-47 in Genetic Contributions to Yield Gains of Five Major Crop Plants, W. R. Fehr, ed. Crop Science Society of Ameica Special Publication Number 7. Madison, Wisconsin: CSSA.

Economic Research Service. 1995. New Crop Varieties: Intellectual Property Rights Spur Development of New Crop Varieties. AREI Updates 14. Washington, D.C.: US Department of Agriculture.

Economic Research Service. 1997. Seed Use, Costs, and Trade. AREI Updates 8. Washington, D.C.: US Department of Agriculture.

Economic Research Service. 1999a. Genetically Engineered Crops for Pest Management. USDA.[Online].Available:http://www.econ.ag.gov/whatsnew/issues/biotech/index.htm [1999, June].

Economic Research Service. 1999b. Value Added and Net Farm Income Data, 1950-1997. Washington, D.C.: US Department of Agriculture. [Online].Available:http://www.econ.ag.gov/briefing/fbe/fi/finfidmu.htm [1999, June].

Eigenbrode, S. D., S. Moodie, and T. Castagnola. 1995. Predators mediate host plant resistance to a phytophagous pest in cabbage with glossy leaf wax. Entomol. Exp. Appl. 77:335-342.

Eleven Scientific Societies. 1996. Appropriate oversight for plants with inherited traits for resistance to pests. A Report from 11 Professional Scientific Societies. Chicago: Institute of Food Technologists. [Online]. Available: www.ift.org/resource/policy/stat_tes/G-061.shtml. [1999, July 8].

Ellstrand, N., and C. Hoffman. 1990. Hybridization as an avenue of escape for engineered genes. Bioscience 40:438-442.

Ellstrand, N. C., and D. R. Elam. 1993. Population genetic consequences of small population size: Implications for plant conservation. Annu. Rev. Ecol. Syst. 24:217-242.

Ellstrand, N. C., and K. W. Foster. 1983. Impact of population structure on the apparent outcrossing rate of grain sorghum (*Sorghum bicolor*). Theor. Appl. Gen. 66:323-327.

Ellstrand, N. C., H. C. Prentice, and J. F. Hancock. 1999. Gene flow and introgression from domestic plants into their wild relatives. Annu. Rev. Ecol. Syst. 30:539-563.

Ehlenfeldt, M. K., and J. P. Helgenson. 1987. Fertility of somatic hybrids from protoplast fusion of *Solanum brevidens* and *S. tuberosum*. Theor. Appl. Gent. 73:395-402.

EPA (US Environmental Protection Agency). 1988a. 7 C.F.R. § 152. Source: 53 Fed. Reg. 15975.

EPA (US Environmental Protection Agency). 1988b. Guidance for the re-registration of pesticide products containing *Bacillus thuringiensis* as the active ingredient. Re-registration Standard 540/RS-89-023. Washington, D.C.: EPA, Office of Pesticides and Toxic Substances.

EPA (US Environmental Protection Agency). 1990. National Survey of Pesticides in Drinking Water Wells: Phase I Report. Publication No. EPA 570/9-90-015. Washington, D.C.: Office of Water and Office of Pesticides and Toxic Substances

EPA (US Environmental Protection Agency). 1994a. Plant-pesticides subject to the Federal Insecticide, Fungicide, and Rodenticide Act; proposed rule. 59 Fed. Reg. 60496.

EPA (US Environmental Protection Agency). 1994b. Watermelon mosaic virus-2 coat protein, zucchini yellow mosaic virus coat protein, and the genetic material necessary for production of these proteins in transgenic squash plants. 59 Fed. Reg. 38149.

EPA (US Environmental Protection Agency). 1994c. Plant-pesticides; Proposed Exemption From the Requirement of a Tolerance Under the Federal Food, Drug, and Cosmetic. 59 Fed. Reg. 60535.

EPA (U.S. Environmental Protection Agency). 1994d. Regulatory Impact Assessment: Proposed Testing and Data Requirements for Registering Pesticides Under the Federal Insecticide, Fungicide, and Rodenticide Act as Amended, 1988. Washington, DC.

EPA (US Environmental Protection Agency). 1995a. Pesticide Fact Sheet: Plant-Pesticide *Bacillus* subsp. *tenebrionis* Delta Endotoxin and its Controlling Sequences in Potato. Issued May 5, 1995.

EPA (US Environmental Protection Agency). 1995b. Pesticide Fact Sheet: *Bacillus thuringiensis* subsp. *kurstaki* Delta Endotoxin and its Controlling Sequences as Expressed in Cotton. Issued October 31, 1995.

EPA (US Environmental Protection Agency). 1995c. Plant pesticide *Bacillus thuringiensis* CryIIIA Delta Endotoxin and the Genetic Material Necessary For its Production; Tolerance Exemption. 60 Fed. Reg. 21725.

EPA (US Environmental Protection Agency). 1995d. Notice of Limited Plant Propagation Registration for a Plant-Pesticide. 60 Fed. Reg. 4910.

EPA (US Environmental Protection Agency). 1996. Plant Pesticides Subject to the Federal Insecticide, Fundicide, and Rodenticide Act and the Federal Food, Drug, and Cosmetic Act; Reopening of Comment Period. 61 Fed. Reg. 37891.

EPA (US Environmental Protection Agency). 1997a. Pesticide Fact Sheet: *Bacillus thuringiensis* Cry IA(b) Delta Endotoxin and the Genetic Material Necessary for its Production in Corn. Issued August 28, 1997.

EPA (US Environmental Protection Agency). 1997b. Plant-pesticides; Supplemental Notice of Proposed Rulemaking. 62 Fed. Reg. 27132.

EPA (US Environmental Protection Agency). 1997c. Proceedings of the Environmental Protection Agency Plant Pesticide Workshop, July 17-18, A. A. Sorenson, ed. Washington, D.C.: Center for Agriculture in the Environment, American Farmland Trust, National Foundation for Integrated Pest Management Education.

EPA (US Environmental Protection Agency). 1998a. Pesticide Fact Sheet: *Bacillus thuringiensis* Cry IA(b) Delta Endotoxin and the Genetic Material Necessary for its Production in Corn. Update to Include Popcorn Use. Issued April 1998.

EPA (US Environmental Protection Agency). 1998b. Pesticide Fact Sheet for *Bacillus thuringiensis* subsp. *kurstaki* CryIAc Delta Endotoxin and the Genetic Material Necessary for its Production in Corn. Issued in August 1998.

EPA (US Environmental Protection Agency). 1998c. Pesticide Fact Sheet: *Bacillus thuringiensis* subspecies *tolworthi* Cry 9C Protein and the Genetic Material Necessary for its Production in Corn. Issued May 1998.

EPA (US Environmental Protection Agency). 1998d. The Environmental Protection Agency's White Paper on Bt Plant-pesticide Resistance Management. Publication No. 739-S-98-001. Washington, D.C.: Biopesticides and Pollution Prevention Division.

EPA (US Environmental Protection Agency). 1999a. EPA Plant Pesticide Regulatory Decisions. [Online]. Available: http://www.aphis.usda.gov/biotech/EPA/index.html [1999, December 2].

EPA (US Environmental Protection Agency). 1999b. Office of Pesticide Programs: Biopesticides Homepage. [Online]. Available: http://www.epa.gov/oppbppd1/biopesticides/index.htm [1999, June].

EPA (US Environmental Protection Agency). 1999c. Plant-Pesticides; Supplemental Notice of Availability of Information. 64 Fed. Reg. 19958.

EPA (US Environmental Protection Agency). 1999d. Office of Pesticide Programs: Pesticide Product Information System. [Online]. Available: http://www.epa.gov/opppmsd1/PPISdata/index.html [1999, June].

EPA (US Environmental Protection Agency). 1999e. Notice to manufacturers, formulators, producers and registrants of pesticide products: Voluntary pesticide resistance management labeling proposal based on target site/mode of action. Draft Pesticide Registration Notice 99-. Washington, D.C.: EPA, Office of Pesticide Programs.

EPA (US Environmental Protection Agency). 1999f. Biopesticide Speech to OECD: Policy and Regulations for New Organisms Supporting Sustainable Pest Management in the United States. [Online]. Available: http://www.epa.gov/oppbppd1/biopesticides/otherdocs/oecd_speech.htm [1999, July 8]. Washington, D.C.: EPA, Office of Pesticide Programs.

EPA (US Environmental Protection Agency). 1999g. Staff observations communicated to the committee. June, 1999.

EPA (US Environmental Protection Agency). 1999h. Letter to Bt corn registrants. December 12. [Online]. Available: http://www.epa.gov/oppbppd1/biopesticides/otherdocs/bt_corn_ltr.htm [2000, February 8]. Washington, D.C.: EPA, Office of Pesticide Programs.

EPA and USDA. 1999. EPA and USDA position paper on insect resistance management in Bt crops. [Online]. Available: http://www.epa.gov/pesticides/biopesticides/otherdocs/bt_position_paper_618.htm [1999, December 2].

Estruch, J. J., N. B. Carozzi, N. Desai, N.B. Duck, G. W. Warren, and M. G. Koziel. 1997. Transgenic plants: An emerging approach to pest control. Nat. Biotechnol. 15:137-141.

Ewen, S.W.B, and A. Pusztai. 1999. Effect of diets containing genetically modified potatoes expressing *Galanthus nivalis* lectin on rat small intestine. Lancet 354(9187):1353-1354.

Falck-Zepeda J. B., G. Traxler, and R. G. Nelson. 1999. Surplus Distribution from the Introduction of a Biotechnology Innovation. Auburn University, LOC: Department of Agricultural Economics and Rural Sociology.

Falk, B. W., and Bruening, G. 1994. Will transgenic crops generate new viruses and new diseases? Science 263:1395-1396.

FAO/WHO. 1996. (Food and Agriculture Organization of the United Nations and World Health Organization). Biotechnology and food safety. Report of a Joint FAO/WHO Consultation. FAO Food and Nutrition Paper No. 61. Rome: FAO.

Farrar, R. R. Jr., and G. Kennedy. 1993. Field cage performance of two tachinid parasitoids of the tomato fruitworm on insect resistant and susceptible tomato lines. Entomol. Exp. Appl. 67:73-78.

FDA (Food and Drug Administration). 1992. Statement of Policy: Foods Derived From New Plant Varieties. 57 Fed. Reg. 22984.

FDA (Food and Drug Administration). 1994a. Secondary Direct Food Additives Permitted in Food for Human Consumption; Food Additives Permitted in Feed and Drinking Water of Animals; Aminoglycoside 3'-Phosphotransferase II. 59 Fed. Reg. 26700.

FDA (Food and Drug Administration). 1994b. Conference on scientific issues related to potential allergenicity in transgenic food crops. Annapolis, MD. April 18-19, 1994.

FDA (Food and Drug Administration). 1997a. National Environmental Policy Act: Revision of Policies and Procedures, Final Rule. 62 Fed. Reg. 40569.

FDA (Food and Drug Administration). 1997b. Supporting Statement for Food Additive Petitions. OMB Approval No. 0910-0016.

FDA (Food and Drug Administration). 1997c. Guidance on Consultation Procedures; Foods Derived From New Plant Varieties. [Online]. Available: http://vm.cfsan.fda.gov/~lrd/consulpr.html [1999, July 21].

FDA (Food and Drug Administration). 1998. Guidance for Industry: Use of Antibiotic Resistance Marker Genes in Transgenic Plants (Draft Guidance). Docket No. 98D-0340. [Online]. Available: http://vm.cfsan.fda.gov/~dms/opa-armg.html [1999, December 2].

FDA (Food and Drug Administration). 1999a. Biotechnology Homepage. Information from the Center for Food Safety and Applied Nutrition and FDA [Online]. Available: http://vm.cfsan.fda.gov/~lrd/biotechm.html [1999, July 21].

FDA (Food and Drug Administration). 1999b. Foods Derived from New Plant Varieties Derived through Recombinant DNA Technology: Final Consultations under FDA's 1992 Policy. [Online]. Available: http://vm.cfsan.fda.gov/~lrd/biocon.html [1999, July 21].

Federici, B. A. 1998. Broadscale use of pest-killing plants to be true test. California Agric. 52:14-20.

Fehr, W. R., ed. 1987. Principles of Cultivar Development, vol. 1. Theory and Technique. New York: Macmillan.

Feitelson, J. S., J. Payne, and L. Kim. 1992. *Bacillus thuringiensis*: Insects and beyond. Bio/technology 10:271-275.

Feldman, J., G. Reed, J. Wyman, J. Stewart, and T. Stone. 1994. Genetically modified Colorado potato beetle resistant potato plants, foliar-applied microbial B.t.t., and conventional insecticides: comparative impacts on non-target arthropods. Public Interest Document. EPA Registration Packet 524-474. Submitted to EPA for Monsanto Company's registration for NatureMark NewLeaf Potato.

Ferguson, J. E., and R. L. Metcalf. 1985. Cucurbitacins: Plant-derived defense compounds for Diabroticites (Coleoptera: Chrysomelidae). J. Chem. Ecol. 11:311-317.

Feyerherm, A. M., K. E. Kemp, and G. M. Paulsen. 1988. Wheat yield analysis in relation to advancing technology in the Midwest United States. Agron. J. 80:998-1001.

Finer, J. J, K.R. Finer, and T Ponappa. 1999. Particle Bombardment Mediated Transformation. Pp. 59-80 in Plant Biotechnology: New Products and Applications, J. Hammond, P. Mcgarvey, and V. Yusibov, eds. New York: Springer-Verlag.

Finney, K. F., W. T. Yamazaki, V. L. Youngs, and G. L. Rubenthaler. 1987. Quality of hard, soft, and durum wheats. Agronomy 13: 677-748.

Fleming, L., J. B. Mann, J. Bean, T. Briggle, J. R. Sanchez-Ramos. 1994. Parkinson's disease and brain levels of organochlorine pesticides. Ann. Neurol. 36:100-103.

Flor, H. H. 1971. Current status of the gene-for-gene concept. Annu. Rev. Phytopathol. 9:275-296.

Fraser, R. S. S. 1990. The genetics of plant-virus interactions: mechanisms controlling host range, resistance and virulence. Pp. 71-92 in Recognition and Response in Plant-Virus Interaction. NATO ASI Series H, v. 41. New York: Springer-Verlag.

Fray, R. G., A. Wallace, P. D. Fraser, D. Valero, P. Hedden, P. M. Bramley, and D. Grierson. 1995. Constitutive expression of a fruit phyoene synthase gene in transgenic tomatoes causes dwarfism by redirecting metabolites from the gibberellin pathway. Plant J. 8:963-701.

Fredshavn J. R., G. S. Poulsen, I. Huybrechts, P. Ruselsheim. 1995. Competitiveness of transgenic oilseed rape. Transgenic Res. 4:142-148.

Free, J. B. 1970. Insect Pollination of Crops. London: Academic Press.

Free, J. B., I. H. Williams, P. C. Longden, M. G. Johnson. 1975. Insect pollination of sugar beet (*Beta vulgaris*) seed crop. Ann. Appl. Biol. 81:127-134.

Frey K. J. 1996. National Plant Breeding Study I: Human and Financial Resources Devoted to Plant Breeding Research and Development in 1994. Special Report 98, Iowa Agriculture and Home Economics Experiment Station. Ames: Iowa State University.

Frey, M., P. Chomet, E. Glawischnig, C. Stettner, S. Grun, A. Winklmair, W. Eisenreich, A. Bacher, R. B. Meeley, S. P. Briggs, K. Simcox, and A. Gierl. 1997. Analysis of a chemical plant defense mechanism in grasses. Science 277:696-699.

Friebe, B. J. Jiang, W. J. Raupp, R. A. MacIntosh, and B. S. Gill. 1996. Characterization of wheat-alien translocations conferring resistance to diseases and pests: Current status. Euphytica 91:59-87.

Friedland J, and S. Kilman. 1999. Produce Market: As Geneticists Develop an Appetite for Greens, Mr. Romo Flourishes. Wall Street Journal. January 28.

Friedman, M., and G. M. McDonald. 1997. Potato glycoalkaloids: chemistry, analysis, safety and plant physiology. Crit. Rev. Plant Sci. 16:55-132.

Fuchs, M., and D. Gonsalves. 1995. Resistance of transgenic hybrid squash ZW-20 expressing the coat protein genes of zucchini yellow mosaic-virus and watermelon mosaic-virus-2 to mixed infections by both potyviruses. Bio/Technology 13:1466-1473.

Fuchs, M., and D. Gonsalves. 1999. Gene flow with virus-resistant commercial transgenic squash. Proceedings of the Fifth International Biotechnology Risk Assessment Symposium. Braunschweig, Germany.

Fuchs, M., F. E. Klas, J. R. McFerson, and D. Gonsalves. 1998. Transgenic melon and squash expressing coat protein genes of aphid-borne viruses do not assist the spread of an aphid-transmissable strain of cucumber mosaic virus in the field. Transgenic Res. 7:449-462.

Fuchs, R. L., J. E. Ream, B. G. Hammond, M. W. Naylor, R. M. Leimgruber, S. A. Berberich. 1993. Safety assessment of the neomycin phosphotransferase II (NPTII) protein. Bio/Technology 11(13):1543-1547.

Fuglie K., N. Ballenger, K. Day, C. Klotz, M. Ollinger, J. Reilly, U. Vasavada, and J. Yee. 1996. Agricultural Research and Development: Public and Private Investments Under Alternative Institutions. Agricultural Economic Report 735, Economic Research Service. Washington, D.C.: US Department of Agriculture.

Genetic Engineering News. 1998. Guide to Biotechnology Companies, 1997. New York: Mary Ann Leibert.

Georghiou, G. P. 1986. The magnitude of the resistance problem. Pp. 14-39 in Pesticide Resistance: Strategies and Tactics for Management. National Research Council. Washington, D.C.: National Academy Press.

Gianessi, L. P., and J. E. Carpenter. 1999. Agricultural Biotechnology: Insect Control Benefits. National Center for Food and Agricultural Policy. [Online]. Available: http://www.bio.org/food&ag/ncfap/ag_bio.htm [1999, July 8].

Gilbert-Albertinin, F., H. Lecoq, M. Pitrat and J. L. Nicolet. 1993. Resistance of *Curcurbita moschata* to watermelon mosaic virus type 2 and its genetic relation to resistance to zucchini yellow mosaic virus. Euphytica 69:231-237.

Gill, S. S., E. A. Cowles, and P. V. Pietrantonio. 1992. The mode of action of *Bacillus thuringiensis* endotoxins. Annu. Rev. Entomol. 37:615-636

Giroux, S., J. C. Cote, C. Vincent, P. Martel, and D. Coderre. 1994. Bacteriological insecticide M-ONE effects on predation efficiency and mortality of adult *Coleomegilla maculata lengi* (Coleoptera: Coccinellidae). J. Econ. Entomol. 87:39-43.

Goldburg, R. J. 1989. USDA's Oversight of Biotechnology. Pp. 137-148 in Reform and Innovation of Science and Education: Planning for the 1990 Farm Bill. U.S. Senate Committee on Agriculture, Nutrition, and Forestry. Washington, D.C.: Government Printing Office.

Goldburg, R. J., and G. Tjaden. 1990. Are B.T.K. plants really safe to eat? Bio/technology 8:1011-1015.

Gonsalves, D. 1998. Control of papaya ringspot virus in papaya: A case study. Annu. Rev. Phytopathol. 36:415-437.

Goolsby, D.A., W. A. Battaglin, G. B. Lawrence, R. S. Artz, B. T. Aulenbach, R. P. Hooper, D. R. Keeney, and G. J. Stensland. 1999. Flux and Sources of Nutrients in the Mississippi-Acthafalaya River Basin, Topic 3 Report. Washington, D.C.: National Ocean Service, US Department of Commerce.

Gould, F. 1978. Resistance of cucumber varieties to *Tetranychus urticae* (*Acaria Tetranychidae*) genetic and environmental determinants. J. Econ. Entomol. 71:680-683.

Gould, F. 1983. Genetics of plant-herbivore systems: interactions between applied and basic study. Pp. 599-653 in Variable Plants and Herbivores in Natural and Managed Systems, R. Denno and B. McClure, eds. New York: Academic Press.

Gould, F. 1986a. Simulation models for predicting durability of insect-resistant germ plasm: A deterministic diploid, two-locus model. Environ. Entomol. 15:1-10.

Gould, F. 1986b. Simulation models for predicting durability of insect-resistant germ plasm: Hessian fly (Diptera: Cecidomyiidae)-resistant winter wheat. Environ. Entomol. 15:11-23.

Gould, F. 1988a. Evolutionary biology and genetically engineered crops. BioScience 38:26-33.

Gould, F. 1988b. Genetics of pairwise and multispecies plant-herbivore coevolution. Pp. 13-55 in Chemical Mediation of Coevolution, K. Spencer, ed. New York: Academic Press.

Gould, F. 1995. Comparisons between resistance management strategies for insects and weeds. Weed Technol. 9:830-839.

Gould, F. 1998. Sustainability of transgenic insecticidal cultivars: Integrating pest genetics and ecology. Annu. Rev. Entomol. 43:701-726.

Gould, F. and B. E. Tabashnik. 1998. Bt-cotton resistance management. Pp. 67-105 in Now or Never: Serious New Plans to Save a Natural Pest Control, M. Mellon and J Rissler, eds. Cambridge, MA: Union of Concerned Scientists.

Gould, F., G. G. Kennedy, and M. T. Johnson. 1991. Effects of natural enemies on the rate of herbivore adaptation to resistant host plants. Ent. Exp. Appl. 58:1-14.

Grant, S. R. 1999. Dissecting the mechanisms of posttranscriptional gene silencing: Divide and conquer. Cell 96:303-306.

Green, G. J., and A. B. Campbell. 1979. Wheat cultivars resistant to *Puccinia graminis tritici* in western Canada: Their development, performance, and economic value. Can. J. Plant Pathol. 1:3-11.

Green, M. B., H. M. LeBaron, and W. K. Moberg, eds. 1990. Managing Resistance to Agrochemicals: From Fundamental Research to Practical Strategies. Washington, D.C.: American Chemical Society.

Greene, A. E., and R. F. Allison. 1994. Recombination between viral RNA and transgenic plant transcripts. Science 263:1423-1425.

Greene, A. E., and R. F. Allison. 1996. Deletions in the 3' untranslated region of cowpea chlorotic mottle virus transgene reduce recovery of recombinant viruses in transgenic plants. Virology 225:231-234.

Gressel, J. 1999. Tandem constructs: preventing the rise of superweeds. Trends Biotechnol. 17:361-366.

Griliches, Z. 1958. Hybrid corn: An exploration in the economics of technological change. Econometrica 25:501-522.

Guretzky, J. A., and S. M. Louda. 1997. Evidence for natural biological control: insects decrease survival and growth of a native thistle. Ecol. Appl. 7:1330-1340.

Hammond, B. G., J. L. Vicini, G. F. Hartnell, M. W. Naylor, C. D. Knight, E. H. Robinson, R. L. Fuchs, and S.R. Padgette. 1996. The feeding value of soybeans fed to rats, chickens, catfish, and dairy cattle is not altered by genetic incorporation of glyphosate tolerance. J. Nutr. 126:717-727.

Hammond-Kosack, K. E., Tang, S. J., Harrision, K., and J. D. G. Jones. 1998. The tomato Cf-9 disease resistance gene functions in tobacco and potato to confer responsiveness to the fungal avirulence gene product Avr9. Plant Cell 10:1251-1266.

Hansen, L. and J. Obrycki. 1999a. Non-target effects of Bt corn pollen on the Monarch butterfly (Lepidoptera: Danaidae). Abstract presented at the North Central Branch Meeting of the Entomological Society of America, Des Moines, Iowa, March 29, 1999.

Hansen, L. and J. Obrycki. 1999b. Non-target effects of Bt corn pollen on the Monarch butterfly (Lepidoptera: Danaidae). Abstract presented at the Annual Meeting of the Entomological Society of America, Atlanta, Georgia, December 12-16, 1999.

Hansen, L., and J. Obrycki. 1999c. Non-target effects of Bt corn pollen on the Monarch butterfly (Lepidoptera: Danaidae) [abstract]. [Online]. Available: http://www.ent.iastate.edu/entsoc/ncb99/prog/abs/D81.html. [2000, January 27]. Ames, Iowa: Iowa State University.

Hanson, C. H., M. W. Pederson, B. Berrang, M. E. Wall,and K. H. Davis Jr. 1973. The saponins in alfalfa cultivars. Pp. 33-52. in Anti-Quality Components of Forages, A. G. Matches, ed. Special Publication No. 4. Madison, WI: Crop Science Society of America.

Hanson, W. D. 1959a. The theoretical distribution of lengths of parental gene blocks in gametes of an F_1 individual. Genetics 44:197-209.

Hanson, W. D. 1959b. Early generation analysis of lengths of heterozygous chromosome segments around a locus held heterozygous with backcrossing or selfing. Genetics 44:833-837.

Hare, J. D. 1992. Effects of plant variation on herbivore-natural enemy interactions. Pp. 278-298 in Plant Resistance to Herbivores and Pathogens: Ecology, Evolution, and Genetics, R.S. Fritz, and E.L. Simms, eds. Chicago: University Chicago Press.

Harrison, L. A., M. R. Bailey, N. W. Naylor, J. E. Ream, B. G. Hammond, D. . Nida, B. L. Burnetter, T. E. Nickson, T. A. Mitsky, M. L. Taylor, R. L. Fuchs, and S. R. Padgette. 1996. The expressed protein in glyphosate-tolerant soybean, 5-enolpyruvylshikimate-3-phosphate synthase from Agrobacterium sp. Strain CP4, is rapidly digested *in vitro* and is not toxic to acutely gavaged mice. J. Nutr. 126(3):728-740.

Harrison, S., L. R. Oliver, and D. Bell. 1977. Control of Texas gourd in soybeans. Proc. South Weed Soc. 30:46.

Hart, K. 1999a. House agriculture subcommittees ask EPA to delay plant pesticide regulation pending hearing March 24. Pesticide Toxic Chem. News, March 11, p. 3.

Hart, K. 1999b. Organic farmers sue EPA over Bt plant pesticide registrations. Pesticide Toxic Chem. News, February 18, pp. 10-11.

Hart, K. 1999c. House Agriculture subcommittees question EPA authority to regulate biotech plants; ask agency to re-open comment period on Plant Pesticide Rule. Pesticide Toxic Chem. News, March 25, p. 3.

Hayenga, M. 1998. Structural change in the biotech seed and chemical industrial complex. AgBioForum 1: 43-55. [Online]. Available: http://www.agbioforum.missouri.edu [1999, December 2].

Hayenga, M., and K. Kimle. 1992. Agricultural Inputs and Processing Industries. Ames: Iowa State University.

Health Canada. 1994. Guidelines for the Safety Assessment of Novel Foods, Volume 1, Preamble and Guidance Scheme for Notification. Food Directorate, Health Protection Branch, Health Canada.

Hellenas, K. E., C. Branzell, H. Johnsson, and P. Slanina. 1995. High levels of glycoalkaloids in the established Swedish potato variety Magnum Bonum. J. Sci. Food Agric. 23:520-523.

Hillbeck, A., M. Baumgartner, P. M. Fried, F. Bigler. 1998a. Effects of transgenic *Bacillus thuringiensis* corn-fed prey on mortality and development time of immature *Chrysoperla carnea* (Neuroptera : Chrysopidae). Environ. Entomol. 27:480-487

Hillbeck, A., W. J. Moar, M. Pusztai-Carey, A. Filippini, and F. Bigler. 1998b. Toxicity of *Bacillus thuringiensis* CryIAb toxin to the predator *Chrysoperla carnea* (Neuroptera: Chrysopidae). Environ. Entomol. 27:1255-1263.

Ho, M. W. 1998. Dangerous liaison—deadly gamble. Pp. 143-148 in Agricultural Biotechnology and Environmental Quality: Gene Escape and Pest Resistance, R. W. Hardy and J. B. Segelken, eds. National Agricultural Biotechnology Council Report No. 10. Ithaca, NY: NABC.

Hodgson, E., and S. Meyer. 1997. Pesticides. Pp. 369-387 in Comprehensive Toxicology, I. G. Sipes, C. A. McQueen and A. J. Gandolfi, series eds., vol. 9, Hepatic and Gastrointestinal Toxicology, R. S. McClusky and D. L. Earnest, volume eds. London: Pergamon Press.

Hokanson, S. C., R. Grumet, and J. F. Hancock. 1997. Effect of border rows and trap/donor ratios on pollen-mediated gene movement. Ecol. Appl. 7:1075-1081.

Holm, L. G., D. L. Plucknett, J. V. Pancho, J. P. Herberger. 1977. The World's Worst Weeds. Distribution and Biology. Honolulu: University of Hawaii Press.

Holm, L. G., J. Doll, E. Holm, J. Pancho, and J. Herberger. 1997 World Weeds: Natural Histories and Distributions. New York: Wiley and Sons.

Horner, J. D., J. R. Gosz, and R. G. Cates. 1988. The role of carbon-based plant secondary metabolites in decomposition in terrestrial ecosystems. Am. Natural. 132:869-883.

Horsch, R. B., J. E. Fry, N. L. Hoffman, D. Eichholts, S. G. Rogers, and R. T. Fraley. 1985. A simple and general method for transferring genes into plants. Science 227:1229-1231.

Hough-Goldstein, J., and C. B. Keil. 1991. Prospects for integrated control of the Colorado potato beetle (Coleoptera: Chrysomelidae) using *Perillus bioculatus* (Hemiptera: Pentatomidae) and various pesticides. J. Econ. Entomol. 84:1645-1651.

Hoxter, K. A., and S. P. Lynn. 1992. Activated *Btk* HD-1 protein: A dietary toxicity study with green lacewing larvae. Submission to EPA for Monsanto Company's registration for Bt corn.

Hoxter, K. A., and G. J. Smith 1993. *B.t.t.* protein: a dietary toxicity study with ladybird beetles (*Hippodamia convergens*). Submitted to EPA for Monsanto Company's registration for NatureMark NewLeaf Potato.

Hoy, C. W., J. Feldman, F. Gould, G. G. Kennedy, G. Reed, and J. A. Wyman. 1998. Naturally occurring biological controls in genetically engineered crops. Pp. 185-205 in Conservation Biological Control, P. Barbosa, ed. New York: Academic Press

Huffman W. E., and R. E. Evenson. 1993. Science for Agriculture: A Long-Term Perspective. Ames: University of Iowa Press.

Hutcheson, S. W. 1998. Current concepts of active defense in plants. Annu. Rev. Phytopathol. 36:59-90.

Huttner, S. L., H. I. Miller, and P. G. LeMaux. 1995. U.S. agricultural biotechnology: status and prospects. Technological Forecasting and Social Change. 50:25-40.

Idaho Statesman. 1998. Genetically engineered potato more than just a "Frankenfood". Concern lingers over promise of "supercrop". November 24, 1998.

Inside EPA. 1999. EPA's biotechnology registration attacked by agriculture industry. Page 12, February 19.

Jach, G., B. Gornhardt, J. Mundy, J. Logemann, E. Pinsdorf, R. Leah. J. Schell, and C. Maas. 1995. Enhanced quantitative resistance against fungal diseases by combinatorial expression of different barley antifungal proteins in transgenic tobacco. Plant J. 8:97-109.

Jackson, D., R. Symons, and P. Berg. 1972. Biochemical method for inserting new genetic information into DNA of simian virus 40: Circular SV40 DNA molecules containing lambda phage genes and the galactose operon of *Escherichia coli*. Proc. Natl. Acad. Sci. USA 69:2904-2909.

Jackson, L. S., E. H. Robinson, D. L. Nida, and P. R. Sanders. 1995. Evaluation of the European corn borer resistant corn line MON 801 as a feed ingredient for catfish. Submitted to EPA for Monsanto Company's registration for Bt corn.

James, C., ed. 1998. Global Review of Commericialized Transgenic Crops: 1998. No. 8. Ithaca, N.Y.: International Service for the Acquisition of Agri-Biotech Applications.

James, C., ed. 1999. Global Review of Commericialized Transgenic Crops: 1999. No. 12. Ithaca, N.Y.: International Service for the Acquisition of Agri-Biotech Applications.

Johal, G. S., and S. P. Briggs. 1992. Reductase activity encoded by the HM1 disease resistance gene in maize. Science 258:985-987.

Johnson, M. T., F. Gould, and G. G. Kennedy. 1997a. Effect of an entomopathogen on adaptation of *Heliothis virescens* populations to transgenic host plants. Entomol. Exp. Appl. 83:121-135.

Johnson, M. T., F. Gould, and G. G. Kennedy. 1997b. Effects of natural enemies on relative fitness of *Heliothis virescens* genotypes adapted and not adapted to resistant host plants. Entomol. Exp. Appl. 82:219-230.

Johnson, R. 1981. Genetic background of durable resistance. Pp. 5-26 in Durable Resistance in Crops, F. Lamberti, J. M. Waller, and N. A. Van der Graaff, eds. NATO Adv. Sci. Inst. Ser. 55. New York: Plenum.

Johnson, R. 1984. A critical analysis of durable resistance. Ann. Rev. Phytopathol. 22:309-330.

Johnson, R., J. Narvaez, G. An, and C. Ryan. 1989. Expression of proteinase inhibitors I and II in transgenic tobacco plants: Effects on natural defense against *Manduca sexta* larvae. Proc. Natl. Acad. Sci. USA 86:9871-9875.

Jones, S. S., T. D. Murray, and R. E. Allen. 1995. Use of alien genes for the development of disease resistance in wheat. Annu. Rev. Phytopathol. 33:429-443.

Just, R. E., and D. L. Hueth. 1997. Multimarket exploitation: the case of biotechnology and chemicals. Am. J. Agric. Econ. 75:936-945.

Kahl, L. S. 1994. Summary of consultation with Calgene, Inc., concerning Flavr Savr tomatoes, FDA Docket No. 91A-0330, memorandum, HFS-206 to HFS-200, May 17th.

Kalaitzandonakes, N. 1997. Mycogen: building a seed company for the twenty-first century. Rev. Agric. Econ. 19:453-462.

Kareiva, P., and R. Sahakian. 1990. Tritrophic effects of a simple architectural mutation in pea plants. Nature 345:433-434.

Kareiva, P., I. M. Parker, and M. Pascual. 1996. Can we use experiments and models in predicting the invasiveness of genetically engineered organisms? Ecology 77:1670-1675.

Kashyap, R. K., G. G. Kennedy, and R. R. Farrar Jr. 1991. Mortality and inhibition of *Helicoverpa zea* egg parasitism rates by *Trichogramma* in relation to trichome/methyl ketone-mediated insect resistance of *Lycopersicon hirsutum* f. *glagratum*, accession P1 134417. J. Chem. Ecol. 17:2381-2395.

Kasschau, K. D., and J. C. Carrington. 1998. A counter-defensive strategy of plant viruses: Suppression of posttranscriptional gene silencing. Cell 95:461-470.

Kauffman, W. G., and R. V. Flanders. 1985. Effects of variably resistant soybean and lima bean cultivars on *Pediobius foveolatus* (Hymenoptera: Eulophidae), a parasitoid of the Mexican bean beetle, *Epilachna varivestis* (Coleoptera: Coccinellidae). Environ. Entomol. 14:678-682.

Kauffman, W. G., and G. G. Kennedy. 1989. Inhibition of *Campoletis sonorensis* parasitism of *Heliothis zea* and of parasitoid development by 2-tridecanone-mediated insect resistance of wild tomato. J. Chem. Ecol. 15:1919-1930.

Kearney, B., and B. J. Staskawicz. 1990. Widespread distribution and fitness contribution of *Xanthomonas campestris* avirulence gene avrBS2. Nature (London) 346:385-386.

Keck, P. J., and S. R. Sims. 1993. Aerobic soil degradation of Colorado potato beetle active protein from *Bacillus thuringiensis* subsp. *tenebrionis*. Submitted to EPA for Monsanto Company's registration for NatureMark NewLeaf Potato.

Keck, P. J., S. R. Sims, and D. E. Bartnicki. 1993. Assessment of the metabolic degradation of the Colorado potato beetle (CPB) active protein in simulated mammalian digestive models. Submitted to EPA for Monsanto Company's registration for NatureMark NewLeaf Potato.

Keeler, R. F., K. R. VanKampen, and L. F. James, eds. 1978. Effects of Poisonous Plants on Livestock. New York: Academic Press.

Kendall, P. 1999. Monarch butterfly so far not imperiled—gene–altered corn gets an early OK in studies. Chicago Tribune, November 2, page 4.

Kennedy, G. G., Gould, F., Deponti, O. M. B., and R. E. Stinner. 1987. Ecological, agricultural, genetic, and commercial considerations in the deployment of insect-resistant germplasm. Environ. Entomol. 16:327-328.

Kilman, S. 1999. In new world of tough plants, pesticide sales soften. Wall Street Journal. 16 June.

Kilpatrick, R. A. 1975. New wheat cultivars and longevity of rust resistance, 1971-1975. Beltsville, MD: US Department of Agriculture, Agricultural Research Service.

Kimber, I, N. I. Kerkvliet, S. L. Taylor, J. D. Astwood, K. Sarlo, and R. J. Dearman. 1999. Toxicology of protein allergenicity: Prediction and characterization. Toxicol. Sci. 48:157-162.

Kirkpatrick, K. J., and H. D. Wilson. 1988. Interspecific gene flow in Cucurbita: *C. texana* vs. *C. pepo*. Am. J. Bot. 75:519-527.

Klein, T. M., E. D. Wolff, R. Wu, and J. C. Sanford. 1987. High-velocity micorprojectiles for delivering nucleic acids into living cells. Nature 327:70-73.

Klinger, T., and N. C. Ellstrand. 1994. Engineered genes in wild populations: Fitness of weed-crop hybrids of *Raphanus sativus*. Ecol. Appl. 4: 117-120.

Klinger, T., P. E. Arriola, and N. C. Ellstrand. 1992. Crop-weed hybridization in radish (*Raphanus sativus*): Effects of distance and population size. Am. J. Bot. 79:1431-1435.

Kloppenburg, Jr., J. R. 1988. First the Seed: The Political Economy of Plant Biotechnology, 1492-2000. New York: Cambridge University Press.

Knott, D. R. 1989. The Wheat Rusts—Breeding for Resistance. Monograph on theoretical and applied genetics 12. New York: Springer-Verlag.

Knutson, R. S., C. R. Taylor, J. B. Penson, Jr., and .E. G. Smith. 1990. Economic Impacts of Reduced Chemical Use. College Station, TX: Knutson and Associates.

Koonin, E. V., and V. V. Dolja. 1993. Evolution and taxonomy of positive-strand RNA viruses: Implications of comparative analysis of amino acid sequences. Crit. Rev. Biochem. Mol. Biol. 28:375-430.

Kramer, K. J., and S. Muthukrishnan. 1997. Insect chitinases: Molecular biology and potential use as biopesticides. Insect Biochem. Mol. Biol. 27:887-900.

Kunkel, T., Q. W. Niu, Y. S. Chan, and N. H. Chua. 1999. Inducible isopentenyl transferase as a high-efficiency marker for plant transformation. Nat. Biotechnol. 17:916-919.

Kuiper, H.A., Noteborn, H.P., and A.C.M. Peijnenburg. 1999. Adequacy of methods for testing the safety of genetically modified foods. Lancet 354(9187): 1315-1316.

Lamberti, F., J. M. Waller, and N. A. Van der Graff. 1981. Durable Resistance in Crops. NATO Adv. Sci. Inst. Ser. 55. New York: Plenum.

Langevin, S. A., K. Clay, and J. B. Grace. 1990. The incidence and effects of hybridization between cultivated rice and its related weed red rice (*Oryza sativa L.*). Evolution 44:1000-1008.

Lannou, C., and C. C. Mundt. 1996. Evolution of a pathogen population in host mixtures: Simple race-complex race competition. Plant Pathol. 45:440-453.

Lannou, C., and C. C. Mundt. 1997. Evolution of a pathogen population in host mixtures: Rate of emergence of complex races. Theor. Appl. Genet. 94:991-999.

Laurila, J., I. Lasko, J. P. T. Valkonen, R. Hiltunen, and E. Pehu. 1996. Formation of parental type and novel glycoalkaloids in somatic hybrids between *Solanum brevidens* and *S. tuberosum*. Plant Sci. 118:145-155.

Lavrik, P. B., D. E. Bartnicki, and S. R. Sims. 1993. Colorado potato beetle active *Bacillus thuringiensis* subsp. *tenebrionis* protein dose formulation, dose confirmation, and dose characterization for albino mice: acute toxicity study (ML-92-407). Submitted to EPA for Monsanto Company's registration for NatureMark NewLeaf Potato.

Lavrik, P. B., D. E. Bartnicki, J. Feldman, B. G. Hammond, P. J. Keck, S. L. Love, M. W. Naylor, G. J. Rogan, S. R. Sims, and R. L. Fuchs. 1995. Safety assessment of potatoes resistant to Colorado potato beetle. Pp. 148-158 in Genetically Modified Foods, Safety Issues. K. H. Engel, G. R. Takeoka and R Teranishi, eds. Washington, D.C.: American Chemical Society.

Layman, P. 1999. Genetically modified foods incite U.K. debate. Chem. Engin. News, page9, February 22.

Lee, T. C., M. Bailey, S. Sims, J. Zeng, C. E. Smith, A. Shariff, L. R. Holden, and P. R. Sanders. 1995. Assessment of the equivalence of the *Bacillus thuringiensis* subsp. *kurstaki* HD-1 protein produced in *Escherichia coli* and the European corn borer resistant corn. Submitted to EPA for Monsanto Company's registration for Bt corn.

Leppik, E. E. 1970. Gene centers of plants as sources of disease resistance. Annu. Rev. Phytopathol. 8:323-344.

Li, Z. K., L. J. Luo, H. W. Mei, A. H. Paterson, X. H. Zhao, D. B. Zhong, Y. P. Wang, X. Q. Yu, L. Zhu, R. Tabien, J. W. Stansel, and C.S. Ying. 1999. A "defeated" rice resistance gene acts as a QTL against a virulent strain of *Xanthomonas oryzae* pv. *oryzae*. Mol. and Gen. Genet. 261:58-63

Lichtenberg, E., and D. Zilberman. 1986. The econometrics of damage control: Why specification matters. Am. J. Agric. Econ. 68:261-273.

Lindbo, J. A., and W. G. Dougherty. 1992. Untranslatable transcripts of the tobacco etch virus coat protein gene can interfere with tobacco etch virus replication in transgenic plants and protoplasts. Virology 189:725-733.

Linder, C. R., I. Taha, G. J. Seiler, A. A. Snow, and L. H. Riesberg. 1998. Long-term introgression of crop genes into wild sunflower populations. Theor. Appl. Gen. 96:339-347.

Lindow, S. W., and N. J. Panopoulos. 1988. Field tests of recombinant ice-*Pseudomonas syringae* for biological frost control in potato. Pp. 121-138 in The Release of Genetically Engineered Micro-Organisms, M. Sussman, G. H. Collins, F. A. Skinner, and D. E. Stewart-Tall, eds. London: Academic.

Line, R. F. 1995. Successes in breeding for and managing durable resistance to wheat rusts. Plant Dis. 79:1254-1255.

Linn, S., and W. Arber. 1968. Host specificity of DNA produced by *Escherichia coli*, X. In vitro restriction of phage fd replicative form. Proc. Natl. Acad. Sci. USA 59:1300-1306.

Liu, Y-B., B. E. Tabashnik, T. J. Dennehy, A. L. Patin, A. C. Bartlett. 1999. Development time and resistance to Bt crops. Nature 400:519.

Loegering, W. Q. 1967. Stem rust of wheat. Agronomy 13:307-317.

Lomonossoff, G. P. 1995. Pathogen derived resistance to plant viruses. Annu. Rev. Phytopathol. 33:323-343.

Losey, J. E., L. S. Raynor, M. E. Carter. 1999. Transgenic pollen harms Monarch larvae. Nature 399:214

Lukaszewski, A. 1990. Frequency of 1RS.1AL and 1RS.1BL translocations in United States wheats. Crop Sci. 30:1151-1153.

MacIntosh S. C., T. B. Stone, S. R. Sims, P. L. Hunst, J. T. Greenplate, P. G. Marrone, F. J. Perlak, D. A. Fischhoff, and R. L. Fuchs. 1990. Specificity and efficacy of purified *Bacillus thuringiensis* proteins against agronomically important insects. J. Invertebr. Pathol. 56:258-266.

Maggi, V. L. 1993a. Evaluation of the dietary effect(s) of purified *B.t.t* protein on honey bee adults. Submitted to EPA for Monsanto Company's registration for NatureMark NewLeaf Potato.

Maggi, V. L. 1993b. Evaluation of the dietary effect(s) of purified *B.t.t* protein on honey bee larvae. Submitted to EPA for Monsanto Company's registration for NatureMark NewLeaf Potato.

Mann, C. 1999. Crop scientists seek a new revolution. Science 283:310-314.

Mansfield, J. W. 1983. Antimicrobial compounds. Pp. 237-265 in Biochemical Plant Pathology, J. A. Callow, ed. Chichester, U.K.: Wiley & Sons.

Martos, A., A. Givovich, and H. M. Niemeyer. 1992. Effect of DIMBOA, an aphid resistance factor in wheat, on the aphid predator *Eriopsis connexa* Germar (Coleoptera: Coccinellidae). J. Chem. Ecol. 18:469-479.

Maryanski, J. H. 1999. Information provided to committee concerning FDA informal consultation process relative to transgenic food plants, May 18.

Matsuda T. 1998. Application of transgenic techniques for hypo-allergenic rice. Pp. 311-314 in the Proceedings From the International Symposium on Novel Foods Regulation in the European Union - Integrity of the Process of Safety Evaluation. Berlin, Germany.

Matten, S. R. 1998. EPA regulation of resistance management for Bt plant-pesticides and conventional pesticides. Resistant Pest Mngt. (IRAC) 10:3-9.

Matten, S. R., P. I. Lewis, G. Tomimatsu, D. W. S. Sutherland, N. Anderson, T. L. Colvin-Snyder. 1996. The U.S. EPA's role in pesticide resistance management. Pp. 243-253 in Molecular Genetics and Evolution of Pesticide Resistance. American Chemical Society symposium series. Washington, D.C.: ACS.

Matthews, R. E. F. 1991. Plant Virology, 3rd edition. San Diego: Academic Press.

Maxam, A. M., and W. Gilbert. 1977. A new method for sequencing DNA. Proc. Natl. Acad. Sci. 74:560-564.

McCartney, H. A., and M. E. Lacey. 1991. Wind dispersal of pollen from crops of oilseed rape (*Brassica napus* L.). J. Agric. Sci. 107:299-305.

McClintock, J. T., C. R. Schaffer, and R. D. Sjoblad. 1995. A comparative review of mammalian toxicity of *Bacillus thuringiensis*-based pesticides. Pestic. Sci. 45:95-105.

McCormick, L. L. 1977. Weed survey - southern states. Southern Weed Science Society, Research Report 30:184-215.

McGaughey, W. H., and M. E. Whalon. 1992. Managing insect resistance to *Bacillus thuringiensis* toxins. Science 258:1451-1455.

McIntosh, R. A., and G. N. Brown. 1997. Anticipatory breeding for resistance to rust diseases in wheat. Annu. Rev. Phytopathol. 35:311-326.

McIntosh, R. A., C. R. Wellings, and R. F. Park. 1995. Wheat Rusts. Boston: Kluwer.

McMullen, N. 1987. Seeds and World Agricultural Progress. Washington, D.C.: National Planning Association.

McVey, D. M. 1990. Reaction of 578 spring spelt wheat accessions to 35 races of wheat stem rust. Crop Sci. 30:1001-1005.

Melanson, D., M. D. Chilton, D. Masters–Moore, and W. S. Chilton. 1997. A deletion in an indole synthase gene is responsible for the DIMBOA–deficient phenotype of bxbx maize. Proc. Natl. Acad. Sci. 94:13345-13350.

Mellon, M. and J. Rissler. 1995. Transgenic crops: USDA data on small-scale tests contribute little on commercial risk assessment. Bio/Technology 13:96.

Meselson, M., and R. Yuan. 1968. DNA restriction enzyme from *E. coli*. Nature 217:1110-1114.

Metcalfe, D., Fuchs, R.L., Townsend, R., Sampson, H.A., Taylor, S.L., and J.R. Fordham. eds. 1996a. Allergenicity of Foods Produced by Genetic Modification; Special Supplement. Critical Reviews in Food Science and Nutrition, Volume36 (Suppl.). Boca Raton, Florida: CRC Press.

Metcalfe, D. D., J. D. Astwood, R. Townsend, H.A. Sampson, S.L. Taylor and R.L. Fuchs. 1996b. Assessment of the allergenic potential of foods derived from genetically engineered crop plants. Critical Reviews in Food Science and Nutrition 36(S):S165-186.

Mikkelsen, T. R., B. Andersen, and R. B. Jørgensen. 1996. The risk of crop transgene spread. Nature 380: 31.

Milewski, E. 1997. Overview of the Environmental Protection Agency's Proposed Plant-pesticide Regulation. . Pp. 11-18 in the Proceedings of the Environmental Protection Agency Plant Pesticide Workshop, July 17-18, A. A. Sorenson, ed. Washington, D.C.: Center for Agriculture in the Environment, American Farmland Trust, National Foundation for Integrated Pest Management Education.

Miller, F. R., and Y. Kebede. 1984. Genetic contributions to yield gains in sorghum, 1950 to 1980. Pp. 1-14 in Genetic Contributions to Yield Gains of Five Major Crop Plants. W. R. Fehr, ed. Crop Science Society of America Special Publication Number 7. Madison, WI: CSSA.

Milligan, S. B., J. Bodeau, J. Yaghoobi, I. Kaloshian, P. Zabel, and V. M. Williamson. 1998. The root knot nematode resistance gene Mi from tomato is a member of the leucine zipper, nucleotide binding, leucine–rich repeat family of plant genes. Plant Cell 10:1307-1319.

Morris, W. F., P. M. Kareiva, and P. L. Raymer. 1994. Do barren zones and pollen traps reduce gene escape from transgenic crops? Ecol. Appl. 2:431-438.

Muehlbauer, J. E., G. J. Specht, M. A. Thomas–Compton, P. E. Staswick, and R. L. Bernard. 1988. Near-isogenic lines—a potential resource in the integration of conventional and molecular marker linkage maps. Crop Sci. 28:729-735.

Mullin, J. W., and J. M. Mills. 1999. Economics of Bollgard versus non-Bollgard cotton in 1998. Pp 958-961 in the proceedings of the Beltwide Cotton Conferences, 1999, vol. 2. Memphis, TN: National Cotton Council of America.

Mundt, C. C. 1990. Probability of mutation to multiple virulence and durability of resistance gene pyramids. Phytopathology 80:221-223.

Munkvold, G. 1998. Disease control with Bt corn. Integrated Crop Management. [Online]. Available: http://www.ipm.iastate.edu/ipm/icm/1998/1-19-1998/btdiscon.html [1999, July 8].

Nap, J-P, J. Bijvoet, and W. J. Stiekema. 1992. Biosafety of kanamycin-resistant transgenic plants. Transgenic Res. 1:239.

NAS (National Academy of Sciences). 1987. Introduction of Recombinant DNA-Engineered Organisms into the Environment: Key Issues. Washington, D.C.: National Academy Press.

NAS (National Academy of Sciences). 1998. Beyond Discovery: The Path from Research to Human Benefit. Designer Seeds. Washington, D.C.: National Academy Press.

NASS (National Agricultural Statistics Service). 1998. United States Summary and State Data. Part 51, Chapter 1 in 1997 Census of Agriculture, Volume 1. Washington, D.C.: US Department of Agriculture.

NASS (National Agricultural Statistics Service). 1999. Census of Agriculture. [Online]. Available: http://www.nass.usda.gov/census [1999, December 2].

NIH (National Institutes of Health). 1976. Recombinant DNA research: Guidelines. 41 Fed. Reg. 27901.

NIH (National Institutes of Health). 1978. Guidlelines for Research Involving Recombinant DNA Molecules. 43 Fed. Reg. 60108.

NIH (National Institutes of Health). 1983. Recombinant DNA Research: Action Under Guidelines. 48 Fed. Reg. 16459.

NOSB (National Organic Standards Board). 1996. Biotechnology Policy. Adopted at Indianapolis, IN. Washington, D.C.: USDA National Organic Program

NRC (National Research Council). 1982. Diet, Nutrition, and Cancer. Washington, D.C.: National Academy Press.

NRC (National Research Council). 1983. Risk Assessment in the Federal Government. Washington, D.C.: National Academy Press.

NRC (National Research Council). 1986. Pesticide resistance: Strategies and tactics for management. Committee on Strategies for the Management of Pesticide Resistant Pest Populations. Washington, D.C.: National Academy Press.

NRC (National Research Council). 1989. Field Testing Genetically Modified Organisms: Framework for Decision. Washington, D.C.: National Academy Press.

NRC (National Research Council). 1996. Ecologically Based Pest Management: New Solutions for a New Century. Washington, D.C.: National Academy Press.

NSTC (National Science Technology Council). 1995. Biotechnology for the 21st Century: New Horizons. A Report from the Biotechnology Research Subcommittee. National Science and Technology Council, Committee on Fundamental Science. Washington, D.C.: US Government Printing Office.

Natural Resources Conservation Service. 1998. National Resources Inventory-1997 State of the Land Update. Washington, D.C.: U.S. Department of Agriculture.

Naylor, M. 1992. Acute oral toxicity study of *Btk* HD-1 tryptic core protein in albino mice. Submitted to EPA for Monsanto Company's registration for Bt corn.

Naylor, M. 1993a. Acute oral toxicity study of *B.t.t* protein in albino mice. Submitted to EPA for Monsanto Company's registration for NatureMark NewLeaf potato.

Naylor, M. 1993b. Acute oral toxicity study of *Bacillus thuiringiensis* var. *kurstaki* [Cry1Ac] HD-73 protein in albino mice. Submitted to EPA for Monsanto Company's registration for Bollgard cotton.

Nester, E. W., M. P. Gordon, R. M. Amasino, and M. F. Yanofsky. 1983. Mutational analysis of the virulence region of an *Agrobacterium tumefaciens* T_i plasmid. J. Bacteriol. 153:878-883.

Nordlee, J. A., S. L. Taylor, J. A. Townsend, L. A. Thomas, and R. K. Bush. 1996. Identification of a Brazil-nut allergen in transgenic soybeans. N. Engl. J. Med. 334:688-694.

Obrycki, J. J., and M. J. Tauber. 1984. Natural enemy activity on glandular pubescent plants in the greenhouse, an unreliable predictor of effects in the field. Environ. Entomol. 13:679-683.

OECD (Organisation for Economic Cooperation and Development). 1993a. Safety Considerations for Biotechnology: Scale-up of Crop Plants. Paris: OECD.

OECD (Organisation for Economic Cooperation and Development). 1993b. Safety Evaluation of Foods Produced by Modern Biotechnology: Concepts and Principles. Paris: OECD.

OECD (Organisation for Economic Cooperation and Development). 1997. Report of the OECD Workshop on the Toxicological and Nutritional Testing of Novel Foods, Aussois, France, March 5-8. SG/ICGB(98)1. Paris: OECD.

Oliver, L. R., S. A. Harrison, and M. McClelland. 1983. Germination of the Texas Gourd (*Cucurbita texana*) and its control in soybean (*Glycine max*). Weed Sci. 31:700-706.

Ollinger, M., and L. Pope. 1995. Strategic research interests, organizational behavior, and the emerging market for products of plant biotechnology. Technological Forecasting and Social Change. 50:55-68.

Osbourn, A. E. 1996. Preformed antimicrobial compounds and plant defense against fungal attack. Plant Cell 8:1821-1831.

Ostlie, K. R., W. D. Hutchison, and R. L. Hellmich. 1997. Bt corn and European corn borer: Long-term success through resistance management. NCR Publication No.602. St. Paul: University of Minnesota.

OSTP (Office of Science and Technology Policy). 1986. Coordinated Framework for Regulation of Biotechnology. 51 Fed. Reg. 23302.

Paavolainen, L., V. Kitunen, and A. Smolander. 1998. Inhibition of nitrification in forest soil by monoterpenes. Plant Soil 205:147-154.

Palm, C.J., K.K. Donegan, D. Harris, and R.J. Seidler. 1994. Quantitation in soil of *Bacillus thwingiensis* var. *kurstaki* delta-endotoxm from transgemc plants. Mol. Ecol. 3:145-151.

Palm, C. J., D. L. Schaller, K. K. Donegan, and R. J. Seidler. 1996. Persistence in soil of transgenic plant produced *Bacillus thuringiensis* var. *kurstaki* delta-endotoxin. Can. J. Microbiol. 42:1258-1262.

Palmer, W. E. 1995. Effects of modern pesticides and farming systems on northern bob-white quail brood ecology. Dissertation, North Carolina State University, Raleigh.

Palmer, W. E., K. M. Puckett, J. R. Anderson, and P. T. Bromley. 1998. Effects of foliar insecticides on survival of northern bobwhite quail chicks. J. Wildlife Mngt. 62:1565-1573.

Parker, I. M., and P. Kareiva. 1996. Assessing the risks of invasion for genetically engi-neered plants: Acceptable evidence and reasonable doubt. Biol. Cons. 78:193-203.

Paton, J. B. 1921. Pollen and pollen enzymes. Am. J. Bot. 8:471-501.

Perlak, F. J., R. W. Deaton, T. A. Armstron, R. L. Fuchs, S. R. Sims, J. Y. Greenplate and D. A Fischolff. 1990. Insect resistant cotton plants. Bio/Technology 8:939-943.

Perlak, F. J., T. B. Stone, Y. M. Muskopf, L. J. Petersen, G. B. Parker, S. A. McPherson, J. Wyman, S. Love, G. Reed, D. Biever. 1993. Genetically improved potatoes: Protection from damage by Colorado potato beetles. Plant Mol. Biol. 22:313-321.

Pilcher C. D., J. J. Obrycki, M. E. Rice, L. C. Lewis. 1997. Preimaginal development, sur-vival, and field abundance of insect predators on transgenic *Bacillus thuringiensis* corn. Environ. Entomol. 26:446-454.

Pimentel, D., L. McLaughlin, A. Zepp, B. Lakitan, T. Kraus, P. Kleinman, F. Vancini, W. J. Roach, E. Graap, W. S. Keeton, and G. Selig. 1991. Environmental and economic impacts of reducing U.S. agricultural pesticide use. In Handbook of Pest Management in Agriculture, Volume I, Second Edition, D. Pimentel ed. Boca Raton, FL: CRC Press.

Powell-Abel, P., R. S. Nelson, B. De, N. Hoffman, S. G. Rogers, R. T. Fraley, and R. N. Beachy. 1986. Delay of disease development in transgenic plants that express the tobacco mosaic virus coat protein gene. Science 232:738-743.

Prakash, C. S. 1999. Activists oppose transgenic plants in India. Information Systems for Biotechnology News Report. [Online]. Available: http://www.nbiap.vt.edu/news/1999/news99.apr.html#apr9902 [1999, December 2].

Price, P. W., C. E. Bouton, B. A. McPheron, J. N. Thompson, and A. E. Weis. 1980. Interac-tions among three trophic levels, influence of plant on interactions between insect herbivores and natural enemies. Annu. Rev. Ecol. Syst. 11:41-65.

Price, W. C. 1940. Comparative host ranges of six plant viruses. Am. J. Bot. 27:530-541.

Provvidenti, R., R. W. Robinson, and H. M. Munger. 1978. Resistance in feral species to six viruses infecting Curcubita. Plant Disease Reporter 62:326-329.

Pruss, G., X. Ge, X. M. Shi, J. C. Carrington, and V. B. Vance. 1997. Plant viral synergism: The potyviral genome encodes a broad-range pathogenicity enhancer that transactivates replication of heterologous viruses. Plant Cell 9:859-868.

Purrington, C. B., and J. Bergelson. 1995. Assessing weediness of transgenic crops: Indus-try plays plant ecologist. Trends Ecol. Evol. 10:340-342.

Qiu, W, and J. W. Moyer. 1999. Tomato spotted wilt tospovirus adapts to the TSWV N gene-derived resistance by genomic reassortment. Phytopathology 89:575-582.

Rabb, R. L., and J. R. Bradley Jr. 1968. The influence of host plants on parasitism of the eggs of the tobacco budworm. J. Econ. Entomol. 61:1249-1252.

Rao, K. V., K. S. Rathore, T. K. Hodges, X. Fu, E. Stoger, D. Sudhakar, S. Williams, P. Christou, M. Bharathi, D. P. Bown, K. S. Powell, J. Spence, A. M. Gatehouse, and J. A. G. JA. 1998. Expression of snowdrop lectin (GNA) in transgenic rice plants confers resistance to rice brown leafhopper. Plant J. 15:469-477.

Ratcliff, F., B. D. Harrison, and D. C. Baulcombe. 1997. A similarity between viral defense and gene silencing in plants. Science 276:1558-1560.

Raybould, A. F., and A. J. Gray. 1998. Will hybrids of genetically modified crops invade natural communities? Trends Ecol. Evol. 9:85-89.

Raynor, G. S., E. C. Ogden and J. V. Hayes, 1972. Dispersion and deposition of corn pollen from experimental sources. Agron. J. 64:420-427.

Ream, J. E. 1994a. Aerobic soil degradation of *Bacillus thuringiensis* var. *kurstaki* HD-73 protein bioactivity. Submitted to EPA for Monsanto Company's registration for Bollgard cotton.

Ream, J. E. 1994b. Assessment of the *In vitro* digestive fate of *Bacillus thuringiensis* subsp. *kurstaki* HD-1 protein. Submitted to EPA for Monsanto Company's registration for Bt corn.

Ream, J. E. 1994c. Assessment of the *In vitro* digestive fate of *Bacillus thuringiensis* var. *kurstaki* HD-1 protein. Submitted to EPA for Monsanto Company's registration for Bollgard cotton.

Rees, M., and Q. Paynter. 1997. Biological control of Scotch broom: Modeling the determinants of abundance and the potential impact of introduced insect herbivores. J. Appl. Ecol. 34:1203-1221.

Reitz, L. P., and B. E. Caldwell. 1974. Breeding for safety in field crops. Pp. 31-44 in The Effect of FDA Regulations (GRAS) on Plant Breeding and Processing, C. H. Hanson, ed. Crop Science Society of America Special Publ. 5. Madison, WI: CSSA.

Richards, A. J. 1986. Plant Breeding Systems. London: Allen and Unwin.

Riggin-Bucci, T. M., and F. Gould. 1997. Impact of intraplot mixtures of toxic and nontoxic plants on population dynamics of diamondback moth (Lepidoptera: Plutellidae) and its natural enemies. J. Econ. Entomol. 90:241-251.

Rissler, J., and M. Mellon. 1996. The Ecological Risks of Engineered Crops. Cambridge, MA: MIT Press.

Robinson, M. 1998. The seed industry and agricultural biotechnology. Pp. 143-148 in Agricultural Biotechnology and Environmental Quality: Gene Escape and Pest Resistance, R. W. Hardy and J. B. Segelken, eds. National Agricultural Biotechnology Council Report 10. Ithaca, N.Y.: NABC.

Roelfs, A. P. 1979. Estimated losses caused by rust in small grain cereals in the United States, 1918-76. US Department of Agriculture, Agricultural Research Service Miscellaneous Publication No.1363. Washington, D.C.: USDA, ARS.

Roelfs, A. P. 1982. Effects of barberry eradication on stem rust in the United States. Plant Dis. 66:177-181.

Rogan, G. J., J. S. Anderson, J. A. McCreary, and P. B. Lavrick. 1993. Determination of the expression levels of *B.t.t.* and NPTII proteins in potato tissue derived from field grown plants. Submitted for to EPA for Monsanto Company's registration for NatureMark NewLeaf potato.

Rosenthal, G. A., and M. R. Berenbaum, eds. 1991. Herbivores: Their Interactions with Secondary Plant Metabolites, Volume 1. San Diego: Academic Press.

Rossi, M., F. L. Goggin, S. B. Milligan, I. Kaloshian, D. E. Ullman, and V. M. Williamson. 1998. The nematode resistance gene Mi of tomato confers resistance against the potato aphid. Proc. Natl. Acad. Sci. USA 95:9750-9754.

Roush, R. T. 1997. Managing resistance to transgenic crops. Pp. 271-294 in Advances in Insect Control: The Role of Transgenic Plants, N. Carozzi, and M. Koziel, eds. London: Taylor & Francis.

Roush, R. T., and B. E. Tabashnik, eds. 1990. Pesticide Resistance in Arthropods. New York: Chapman & Hall.

Rowell, J .B. 1985. Evaluation of chemicals for rust control. Pp. 561-589 in The Cereal Rusts: Diseases, Distribution, Epidemiology, and Control, vol. 2, A. P. Roelfs, and W. R. Bushnell, eds. Orlando, FL: Academic Press.

Royal Society. 1999. Review of data on possible toxicity of GM potatoes. [Online]. Available: http://www.royalsoc.ac.uk/st_pol54.htm [1999, December 2].

Ryals, J. A., U. H. Neuenschwander, M. G. Willits, A. Molina, H. Y. Steiner, and M. D. Hunt. 1996. Systemic acquired resistance. Plant Cell 8:1809-1819.

Ryan, C. A. 1990. Protease inhibitors in plants: Genes for improving defenses against insects and pathogens. Annu. Rev. Phytopathol. 28:425-449.

Salzman, R. A., I. Tikhonova, B. P. Bordelon, P. M. Hasegawa, and R. A. Bressan. 1998. Coordinate accumulation of antifungal proteins and hexoses constitutes a developmentally controlled defense response during fruit ripening in grape. Plant Physiol. 117:465-472.

Sammons, R. D. 1994a. Assessment of equivalency between E. coli-produced and cotton-produced B.t.k. HD-73 protein and characterization of the cotton-produced B.t.k. HD-73 protein. Submitted to EPA for Monsanto Company's registration for Bollgard cotton.

Sammons, R. D. 1994b. B.t.k. HD-73 protein dose formulation and determination of dose for an acute mouse feeding study ML 92-493. Submitted to EPA for Monsanto Company's registration for Bollgard cotton.

Sanford, J. C., and S. A. Johnston. 1985. The concept of parasite-derived resistance. Deriving resistance genes from the parasite's own genome. J. Theor. Biol. 113:395-405.

Sanger, F., and A. R. Coulson. 1975. A rapid method for determining sequences in DNA by primed synthesis with DNA polymerase. J. Mol. Biol. 94:444-448.

Schafer, J. F. 1987. Rusts, smuts, and powdery mildew. Agronomy 13:542-584.

Scheid, O. M. 1999. Plant viruses. New tools for Swiss army knife. Nature 297:25.

Schimel, J. P., K. VanCleve, R. G. Cates, T. P. Clausen, and P. B. Reichardt. 1996. Effects of balsam poplar (Populus balsamifera) tannins and low molecular weight phenolics on microbial activity in taiga flood plain soil; implications for changes in N cycling during succession. Can. J. Bot. 74:84-90.

Schmidt, J. W. 1984. Genetic contributions to yield gains in weat. Pp. 89-101 in Genetic Contributions to Yield Gains of Five Major Crop Plants , W. R. Fehr, ed. Crop Science Society of America Special Publication Number 7, Madison, WI: CSSA.

Schultheis, J. R., and S. A. Walters. 1998. Yield and virus resistance of summer squash cultivars and breeding lines in North Carolina. HortTechnology 8:31-39.

SAP (Scientific Advisory Panel). 1994. Final Report of the joint FIFRA scientific advisory panel and biotechnology science advisory committee meeting, January 21, 1994. Washington, D.C.: US Environmental Protection Agency.

SAP (Scientific Advisory Panel). 1998. Final report of the FIFRA Scientific Advisory Panel Subpanel on Bacillus thuringiensis (Bt) plant-pesticides and resistance management, February 9 and 10, 1998. Washington, D.C.: US Environmental Protection Agency.

Seefeldt, S. S., R. Zemetra, F. L. Young, and S. S. Jones. 1998. Production of herbicide-resistant jointed goatgrass (Aegilops cylindrica) x wheat (Triticum aestivum) hybrids in the field by natural hybridization. Weed Sci. 46:632-634.

Seiler, G. J. 1992: Utilization of wild sunflower species for the improvement of cultivated sunflower. Field Crops Res. 30:195-230.

Senti, F. R., and R. L. Rizek. 1974. An overview on GRAS regulations and their effect from the viewpoint of nutrition. Pp. 7-20 in The Effect of FDA Regulations (GRAS) on Plant Breeding and Processing. Crop Science Society of America, Special Publication No. 5. Madison, WI: CSSA.

Sharma, H. C., and B. S. Gill. 1983. Current status of wide hybridization in wheat. Euphytica 32:17-31.

Shaver, T. N., and M. J. Lukefahr. 1969. Effect of flavonoid pigments and gossypol on growth and development of bollworm, tobacco budworm, and pink bollworm. J. Econ. Entomol. 62:643-646.

Shechner, D. 1997. Genes and genomics, people and plants: The Arabidopsis sequencing project. Harbor Transcript. 15(3/4):1-9. [Online]. Available: http://www.cshl.org/public/HT/Arabid.html [1999, December 2].

Shelton, A. M., and J. A. Wyman. 1991. Insecticide resistance of diamondback moth (Lepidoptera: Plutellidae) in North America. Pp. 447-454 in Diamondback Moth and Other Crucifer Pests: Proceedings of the Second International Workshop, N. S. Taleker, ed. Taiwan: AVRDC.

Sims, S. R. 1993. Sensitivity of selected insect species to the Colorado potato beetle active protein from *Bacillus thuringiensis* subsp. *tenebrionis*. Submitted to EPA for Monsanto Company's registration for NatureMark NewLeaf Potato.

Sims, S. R. 1994. Sensitivity of insect species to the purified Cry1A(c) insecticidal protein from *Bacillus thuringiensis* var. *kurstaki* (*B.t.k.* HD-73). Submitted to EPA for Monsanto Company's registration for Bollgard cotton.

Sims, S. R., and P. R. Sanders. 1995. Aerobic soil degradiaiton of *Bacillus thuringiensis* var. *kurstaki* HD-1 protein. Submitted to EPA for Monsanto Company's registration for Bt corn.

Sinden, S. L. and R. E. Webb. 1972. Effect of variety and location on the glycoalkaloid content of potatoes. American Potato Journal 49: 334-338.

Small, E. 1984. Hybridization in the domesticated-weed-wild complex. Pp. 195-210 in Plant Biosystematics, W. F. Grant, ed. Toronto: Academic Press

Smith, R. A., R. B. Alexander, and K. J. Lanfear. 1991. Stream water Quality in the Conterminous United States—Status and Trends of Selected Indicators During the 1980's. National Water Summary 1990-1991. US Geological Survey Water Supply Paper 2400. Reston, VA: USGS.

Smith, R. F., and R. Van denBosch. 1967. Integrated control. Pp. 295-340 in Pest Control: Biological, Physical, and Selected Chemical Methods. New York: Academic Press.

Snow, A. A., and P. Morán–Palma. 1997. Commercialization of transgenic plants: potential ecological risks. BioScience 47:86-97.

Snow, A. A., B. Anderson, and R. B. Jørgensen. 1999. Costs of transgenic herbicide resisitance introgressed from *Brassica napus* into weedy *Brassica rapa*. Mol. Ecol. 8:605-615.

Snow, A. A., P. Morán–Palma, L. H. Rieseberg, and A. Wszelaki. 1998. Fecundity, phenology, and seed dormancy of F_1 wild-crop hybrids in sunflower (*Helianthus annuus*, Asteraceae). Am. J. Bot. 85:794-801.

Spalding, R. F., and M. E. Exner. 1993. Occurrence of ntrate in goundwater—A review. J. Environ. Qual. 22:392-402.

Specht, J. E., and J. H. Williams. 1984. Contributions of Genetic Technology to Soybean Productivity—Retrospect and Prospect. Pp. 49-74 in Genetic Contributions to Yield Gains of Five Major Crop Plants, W.R. Fehr, ed. Crop Science Society of America Special Publication No. 7, Madison, WI: CSSA

Stern, V. M., R. F. Smith, R. van den Bosch, and K. S. Hagen. 1959. The integrated control concept. Hilgardia 29:81-101.

Stevenson, G. C. 1965. Genetics and Breeding of Sugar Cane. London: Longmans.

Stewart, C. N., and C. S. Prakesh. 1998. Chloroplast-transgenic plants are not a gene flow panacea. Nat. Biotechnol. 16:401.

Stewart, C. N., J. N. All, P. L. Raymer, and S. Ramachadran. 1997. Increased fitness of transgenic insecticidal rapeseed under insect selection pressure. Mol. Ecol. 6:773-779.

Stoskopf, N. C., D. T. Tomes, and B. R . Christie. 1993. Plant Breeding: Theory and Practice. Boulder, CO: Westview Press.

Sturckow, B., and I. Low. 1961. The effects of some *Solanum* glycoalkaloids on the potato beetle. Entomol. Exp. Appl. 4:133-142.

Tabashnik, B. E. 1994. Evolution of resistance to *Bacillus thuringiensis*. Annu. Rev. Entomol. 39:47-79.

Tallamy, D. W., D. P. Whittington, F. Defurio, D. A. Fontaine, P. M. Gorski, and P. W. Gothro. 1998. Sequestered cucurbitacins and pathogenicity of *Metarhizium anisopliae* (Moniliales: Moniliaceae) on spotted cucumber beetle eggs and larvae (Coleoptera: Chrysomelidae). Environ. Entomol. 27:366-372.

Thompson, C. E., G. Squire, G. R. Mackay, J. E. Bradshaw, J. Crawford, and G. Ramsay. 1999. Regional patterns of gene flow and its consequences for GM oilseed rape. Pages 95-100 in Gene Flow and Agriculture: Relevance for Transgenic Crops, Symposium chaired by P. J. W. Lutman. British Crop Protection Council, Symposium Proceedings No. 72. Farnham: BCPC.

Thorpe, K.W., and P. Barbosa. 1986. Effects of consumption of high and low nicotine tobacco by *Manduca sexta* (Lepidoptera: Sphingidae) on survival of gregarious endoparasitoid *Cotesia congregata* (Hymenoptera: Braconidae). J. Chem. Ecol. 12:1329-1337.

Tiedje, J. M., R. K. Colwell, Y. L. Grossman, R. E. Hodson, R. E. Lenski, R. N. Mack, and R. J. Regal. 1989. The planned introduction of genetically engineered organisms: Ecological considerations and recommendations. Ecology 70:298-315.

Tricoli, D. M., K. J. Carney, P. F. Russell, J. Russell McMaster, D. W. Groff, K. C. Hadden, P. T. Himmel, J. P. Hubbard, M. L. Boeshore, and H. D. Quemada. 1995. Field evaluation of transgenic squash containing single or multiple virus coat protein constructs for resistance to cucumber mosaic virus, watermelon mosaic virus 2, and zucchini yellow mosaic virus. Bio/technology 13:1458-1465.

Trimble, S. W. 1999. Decreased rates of alluvial sediment storage in the Coon Creek River Basin, Wisconsin, 1975-1993. Science 285:1244-1246.

Trumble, J. T., W. Dercks, C. F. Quiros, and R. C. Beier. 1990. Host plant resistance and linear furanocoumarin content of Apium accessions. J. Econ. Entomol. 83:519-525.

Trumble, J. T., M. M. Diawara, and C. F. Quiros. In Press. Breeding resistance in *Apium graveolens* to *Liriomyza trifolii*: antibiosis and linear furanocoumarin content. Acta Horticulturae.

UCS (Union of Concerned Scientists). 1998. Now or Never: Serious New Plans to Save a Natural Pest Control, M. Mellon and J. Rissler, eds. Cambridge, MA: UCS.

US Congress. 1947. Federal Insecticide, Fungicide, and Rodenticide Act , 7 U.S.C. § 136, as amended.

US Congress. 1957. Federal Plant Pest Act, 7 U.S.C. § 150aa-jj, as amended.

US Congress. 1958. Federal Food, Drug, and Cosmetic Act, 21 U.S.C. § 301, as amended.

US Congress. 1969. National Environmental Policy Act, 42 U.S.C. § 4321, as amended.

US Congress. 1976a. Consumer Product Safety Act, 15 U.S.C. § 2501, as amended.

US Congress. 1976b. Toxic Substances Control Act, 15 U.S.C. § 2601, as amended.

US Congress. 1983. Environmental implications of genetic engineering. 98th Congress, 1st session. Hearing Before the Subcommittee on Investigations and Oversight and the Subcommittee on Science, Research and Technology. Committee on Science and Technology. Washington, D.C.: Government Printing Office.

USDA (US Department of Agriculture). 1953. Agricultural Statistics. Washington, D.C.: Government Printing Office.

USDA (US Department of Agriculture). 1987. 7 C.F.R § 340. Introduction of Organisms and Products Altered or Produced Through Genetic Engineering Which are Plant Pests or Which There is Reason to Believe are Plant Pests. Source: 52 Fed. Reg. 22892.

USDA (US Department of Agriculture). 1993. Genetically engineered organisms and products; notification procedures for the introduction of certain regulated articles; and petition for nonregulated status. 58 Fed. Reg. 17044.

USDA (US Department of Agriculture). 1994a. 7 C.F.R. § 201.76. Federal Seed Act Regulations; Minimum Land, Isolation, Field, and Seed Standards. Source: 59 Fed. Reg. 64516. Washington, D.C.: USDA Agricultural Marketing Service.

USDA (US Department of Agriculture). 1994b. Response to Upjohn Company/Asgrow Seed Company Petition 92-204-01 for Determination of Nonregulated Status for ZW-20 Squash. Washington, D.C.: USDA, Animal and Plant Health Inspection Service.

USDA (US Department of Agriculture). 1995a. Environmental Assessment and Finding of No Significant Impact. Petition 94-257-01 for Determination of Nonregulated Status for Colorado Potato Beetle-Resistant Potato Lines BT6, BT10, BT12, BT16, BT17, BT18, and BT23. March 2. Washington, D.C.: USDA Animal and Plant Health Inspection Service.

USDA (US Department of Agriculture). 1995b. 7 C.F.R. § 372, National Environmental Policy Act Implementing Procedures. Source: 60 Fed. Reg. 6002.

USDA (US Department of Agriculture). 1996a. Guide for preparing and submitting a petition for genetically engineered plants. Washington, D.C.: USDA Animal and Plant Health Inspection Service.

USDA (US Department of Agriculture). 1996b. Response to the Asgrow Seed Company Petition 95-352-01 for Determination of Nonregulated Status for CZW-3 Squash. Washington, D.C.: USDA Animal and Plant Health Inspection Service.

USDA (US Department of Agriculture). 1999a. Agricultural Statistics, 1998. Washington, D.C: Government Printing Office.

USDA (US Department of Agriculture). 1999b. Crop lines no longer regulated by USDA. [Online]. Available: http://www.aphis.usda.gov/biotech/not_reg.html [1999, June].

USDA (US Department of Agriculture). 1999c. Field Test Releases in the U.S. [Online]. Available: http://www.nbiap.vt.edu/cfdocs/fieldtests1.cfm. [1999, July]

USDA (US Department of Agriculture). 1999d. Genetically engineered crops for pest management. [Online]. Available: http://www.econ.ag.gov/whatsnew/issues/biotech/index.htm [1999, July 21].

USDA (US Department of Agriculture). 1999e. Regulatory Oversight in Biotechnology: Responsible Agencies – Overview. [Online]. Available: http://www.aphis.usda.gov/biotech/OECD/usregs.htm [1999, July] and http://www.aphis.usda.gov/biotechnology/index.html [1999, December]

USDA (US Department of Agriculture). 1999f. Table for Environmental Releases in the U.S. [Online]. Available: http://www.nbiap.vt.edu/cfdocs/ISBtables.cfm), [1999, June 21]

USDA (US Department of Agriculture). 1999g. Small Grain Losses Due to Rust. [Online]. Available: http:// www.cdl.umn.edu/Loss/Loss.html. [1999, July 8]. St. Paul, MN: USDA Cereal Disease Laboratory.

US Department of Commerce. 1952. U.S. Census of Agriculture, 1950, Volume II: General Report. Washington, D.C.: Government Printing Office.

US Senate. 1984. The potential environmental consequences of genetic engineering. 98th Congress, 2d session. September 25 and 27. Hearings Before the Subcommittee on Toxic Substances and Environmental Oversight. Committee on Environment and Public Works. Washington, D.C.: Government Printing Office.

US Wheat & Barley Scab Initiative. 1998. Scab: A threat to the nation's food system. Scab Initiative News 1:1-2. [Online] Available: http://www.smallgrains.org/scabnews/scab.htm [1999, July 8].

University of Hawaii at Manoa. 1997. Safety Assessment of Ringspot Virus Resistant Papaya Lines 55-1 and 63-1. FDA Biotechnology Notification File 000042 submitted for FDA consultation. January 3, 1997.

Urwin, P. E., C. J. Lilley, M. J. McPherson, and H. J. Atkinson. 1997. Resistance to both cyst and root knot nematodes conferred by transgenic *Arabidopsis* expressing a modified plant cystatin. Plant J. 12:455-461.

Van der Plank, J. E. 1963. Plant Diseases: Epidemics and Control. New York: Academic Press.

Van Raamsdonk, L. W. D., and H. J. Schouten. 1997. Gene flow and establishment of transgenes in natural plant populations. Acta Bot. Nederlandica 46:69-84.

Viscusi, W. K., J. M. Vernon, and J. E. Harrington, Jr. .1995. Economics of Regulation and Antitrust. Cambridge, MA: MIT Press.

Vos, P., G. Simons, T. Jesse, J. Wijbrandi, L. Heinen, R. Hogers, A. Frijters, J. Groenendijk, P. Diergaarde, M. Reijans, J. Fierens–Onstenk, M. deBoth, J. Peleman, T. Liharska, J. Hontelez, and . 1998. The tomato Mi-1 gene confers resistance to both root-knot nematodes and potato aphids. Nat. Biotechnol. 16:1365-1369.

Wahl, I., Y. Anikster, J. Manisterski, and A. Segal. 1984. Evolution at the center of origin. Pp. 39-77 in The Cereal Rusts: Origins, Specificity, Structure, and Physiology, Vol. I, A. P. Roelfs, and W. R. Bushnell, eds. Orlando, FL: Academic Press.

Waloff, N., and O. W. Richards. 1977. The effect of insect fauna on growth, mortality, and natality of broom, *Sarothamnus scoparius*. J. Appl. Ecol. 14:787-798.

Walton, J. D. 1996. Host selective toxins: Agents of compatibility. Plant Cell 8:1723-1733.

Wassenaar, L. I., and K. A. Hobson. 1998. Natal origins of migratory monarch butterflies at wintering colonies in Mexico: New isotopic evidence. Proc. Natl. Acad. Sci. U.S.A. 95:15436-15439.

Weiss, R. 1999. Gene-altered corn's impact reassessed. The Washington Post, November 3, page A3.

Weiss, R. 2000. EPA restricts gene-altered corn in response to concerns. The Washington Post, January 16, page A2.

Whitham, S., S. McCormick, and B. Baker. 1996. The N gene of tobacco confers resistance to tobacco mosaic virus in transgenic tomato. Proc. Natl. Acad. Sci. USA 93:8776-8781.

Whitton, J., D. E. Wold, D. M. Arias, A. A. Snow, and L. H. Rieseberg. 1997. The persistence of cultivar alleles in wild populations of sunflowers five generations after hybridization. Theor. Appl. Genet. 95:33-40.

Williams, M., R. F. Barnes, and J. M. Cassady. 1971. Characterization of alkaloids in palatable and unpalatable clones of *Phalaris arundinacea* L. Crop Sci. 11:213-217.

Williams, M. R. 1997. Cotton insect losses—1979 to 1996. Pp. 854-858 in Proceedings of the Beltwide Cotton Conference, P. Dugger and D.A. Richter, eds. Memphis, TN: National Cotton Council of America.

Williams, M. R. 1998. Cotton insect losses—1997. Pp. 904-925 in Proceedings of the Beltwide Cotton Conference, P. Dugger and D.A. Richter, eds. Memphis, TN: National Cotton Council of America.

Williams, M. R. 1999. Cotton insect losses—1998. Pp. 785-809 in Proceedings of the Beltwide Cotton Conference, P. Dugger and D.A. Richter, eds. Memphis, TN: National Cotton Council of America.

Wilson, H. D. 1990. Gene flow in squash species. BioScience 40:449-455.

Wilson, H. D. 1993. Free-living *Cucurbita pepo* in the United States: viral resistance, gene flow, and risk asessment. Prepared for USDA Animal and Plant Health Inspection. Hyattsville, MD: USDA - APHIS.

Wintermantel, W. M., and J. E. Schoelz. 1996. Isolation of recombinant viruses between cauliflower mosaic virus and a viral gene in transgenic plants under conditions of moderate selection. Virology 223:156-164.

Wrubel, R. P., S. Krimsky, and R. E. Wetzler. 1992. Field testing transgenic plants. BioScience 42:280-289.

Yoon, C.K. 1999. No Consensus on the Effects of Engineering on Corn Crops. November 4. The New York Times.

Zeigler, R. S. 1998. Recombination in *Magnaporthe grisea*. Annu. Rev. Phytopathol. 36:249-275.

Zeller, F. J., and S. L. K. Hsam. 1983. Broadening the genetic variability of cultivated wheat by utilizing rye chromatin. Pp. 161-173 in Proceedings of the Sixth International Wheat Genetics Symposium, S. Sakamoto, ed. Kyoto: Plant Germ-Plasm Institute, Kyoto University.

Zemetra, R. S., J. Hansen, and C. A. Mallory–Smith. 1998. Potential for gene transfer between wheat (*Triticum aestivum*) and jointed goatgrass (*Aegilops cylindrica*). Weed Sci. 46:313-317.

Zeven, A. C., D. R. Knott, and R. Johnson. 1983. Investigation of linkage drag in near isogenic lines of wheat by testing for seedling reaction to races of stem rust, leaf rust, and yellow rust. Euphytica 32:319-327.

Zhu, Q., E. A. Maher, S. Masoud, R. A. Dixon, and C. J. Lamb. 1994. Enhanced protection against fungal attack by constitutive co-expression of chitinase and glucanase genes in transgenic tobacco. Bio/technology 12:807-812.

Zitnak, A., and G. R. Johnston. 1970. Glycoalkaloid content of B5141-6 potatoes. Am. Potato J. 47:256-260.

Appendixes

Appendix A[1]

Costs of Regulating Transgenic Pest-Protected Plants

Erik Lichtenberg, University of Maryland

A.1 INTRODUCTION

The bulk of this report is devoted to the potential risks posed by transgenic pest protected plants and the ways that regulation can mitigate those risks. In other words, this report focuses primarily on the benefits of regulating transgenic pest protected plants, even though those benefits are presented in neither quantitative terms (magnitudes of risk and risk reduction) nor economic ones (the public's willingness to pay for reduction of these risks, increases in sales due to allayed fears about safety, etc.). Yet regulation is desirable only if its benefits outweigh its costs; the mere existence of risk does not imply that regulation is necessary or desirable. This appendix considers potential costs of regulating transgenic pest protected plants and provides evidence regarding the potential magnitudes of some of those costs. Two forms of regulation are considered. One involves regulating transgenic pest protected plants as pesticides under the Federal Insecticide, Rodenticide, and Fungicide Act (FIFRA). The other involves regulating environmental effects under the Federal Plant Pest Act (FPPA) and related legislation and regulating food safety under the Federal Food, Drug, and Cosmetic Act (FFDCA) administered

---

[1]This appendix was authored by an individual committee member and is not part of the committee's consensus report. The committee as a whole may not necessarily agree with all of the contents of appendix A.

211

by the Food and Drug Administration (FDA)—as would occur under the Coordinated Framework if transgenic pest protected plants were not legally categorized as pesticides.

Regulation of transgenic pest-protected plant products will affect the crop-protection and seed industries. It may also affect crop breeding more broadly with implications for future agricultural productivity growth and for the structure of US agriculture. The possibility that regulation would function as a barrier to the entry of new participants is of special concern in this regard. The notion that regulation can function in such a manner is well established in economics. Regulatory requirements can create or enhance economies of scale by increasing the fixed costs of doing business in an industry, thereby limiting entry. Some have gone so far as to argue that established firms may even seek regulation precisely to inhibit entry and thus maintain market power (for a brief discussion see Viscusi et al. 1995).

The costs of generating test data on transgenic pest-protected plants might make regulatory review serve as a barrier to entry. Strict pre-market review of potential products with extensive testing requirements can be expensive. Imposing such fixed costs on newly developed products can have two kinds of negative effects. First, it can increase the potential size of the market (expected sales) needed to break even and thus justify investment in a new plant variety. Regulation could thus be one factor limiting investment in transgenic pest-protected plants with small potential markets; in other words, high testing costs could engender an "orphan crop" problem by discouraging research and development (R&D) aimed at crops for which annual seed or propagule purchases would be small. Second, it can limit entry into the market by entities that have relatively little capital, including small biotechnology startup companies, small to medium seed companies, and public-sector breeders. Limiting entry can reduce competition in varietal development, which in turn can lead to lower overall levels of investment in breeding R&D and affect the future growth and sustainability of agricultural productivity.

Regulation of transgenic pest-protected plant products as pesticides raises special concerns. The Environmental Protection Agency's (EPA) approach to regulation of these products typically involves requiring an applicant to submit more and more-expensive test data than submitted to the US Department of Agriculture (USDA) and the Food and Drug Administration (FDA) during their reviews of comparable products. As a result, regulation of transgenic pest-protected plant products by EPA under FIFRA and FFDCA has the potential to impose more substantial barriers to entry than regulation by USDA and FDA.

This appendix examines the extent to which regulation of transgenic pest-protected plant products has the potential to discourage R&D related to crops that have small seed markets and to create barriers to the entry of less well-capitalized entities. I begin with the background needed to assess that potential. First, I review the roles of plant breeding and pest management in the growth of US agricultural productivity since World War II. I then review the current state of plant-breeding and crop-protection R&D in the United States, both generally and with respect to transgenic pest-protected crops, with an emphasis on the division of labor between the public and private sectors. Third, I review the structure of the seed and agrichemical industries involved in plant-breeding and crop-protection R&D. The seed and agrichemical industries have experienced a wave of merger and acquisition activity in recent years, raising concern about potential adverse effects of concentration on R&D; I examine the structure of the industries in light of the recent changes and discuss what is known about the impacts of transgenic pest-protected plants on competition.

I then consider the potential costs of pre-market regulatory review, with special attention to likely testing requirements for pesticides under FIFRA and FFDCA relative to data and information typically submitted to USDA under the Federal Plant Pest Act and to FDA under FFDCA. I compare those costs to the fixed costs of breeding and estimate the effect of testing requirements on the market size needed to justify investment.

A.2 AGRICULTURAL PRODUCTIVITY IN THE UNITED STATES

Pest management and crop breeding have played important roles in helping US agriculture to maintain impressive growth in productivity over the last 50 years. Between 1948 and 1994, agricultural productivity increased at an average annual rate of more than 1.9%, almost twice the rate of growth of productivity in the US nonfarm economy (Ball et al. 1994). Agricultural productivity growth has continued unabated in the 1990s. Between 1989 and 1994, agricultural productivity grew at an average rate of 2.9%.

Three broad groups have generally shared in the gains from that productivity growth: consumers; input suppliers, processors, and marketers of agricultural products; and farmers.

Consumers in the United States and abroad have been major beneficiaries of productivity growth. From 1948 to 1994, agricultural productivity grew faster than the US population (which increased at an average annual rate of 1.2%), allowing for growth in domestic per capita food consumption and exports simultaneously. US consumer benefits have come partly in the form of increased food consumption and partly in the form of lower food prices (which permit increased consumption of other goods and services). Overall, real per capita expenditures on food increased by 18% from 1948 to 1997, while food prices were declining by roughly 10% relative to the overall price level, as indicated by the fall in the consumer price index (CPI) for food from 108% of the overall CPI in 1948 to 98% in 1997. Real per capita food consumption, as measured by expenditures deflated by the CPI for food, increased by about 31% over that period. Moreover, spending on food has made up a continuously falling share of consumption and income. In 1948, food accounted for 31% of personal consumption expenditures in the United States; by 1997, its share had fallen to 14% (Council of Economic Advisers 1999).

The extent to which input suppliers, processors, and marketers of agricultural products have appropriated gains from increased farm productivity has not been studied. Purchased inputs have played an increasingly important role in agriculture over the last 50 years, as indicated by the increase in the share of the gross value of farm output accounted for by purchased input expenditures, from 18.3% in 1950 to 25.7% in 1997 (Economic Research Service 1999b). The role of processors and marketers in the food industry has increased similarly. In 1948, the farm sector accounted for 41% of the value of food products in the United States; by 1997, that share had fallen to 21%.

The gains from increased agricultural productivity accruing to farmers are reflected in changes in the net value added by the farm sector, which rose in nominal terms from $18.3 billion in 1950 to $92.8 billion in 1997. The average value added per farm in real (1997 dollar) terms more than doubled during that period, increasing from $22,606 in 1950 to $45,565 in 1997 (Economic Research Service 1999b). The average household income of farm operators rose during the period from 60% of the national average to 105% of the national average.

Broadly speaking, the new agricultural technologies underlying productivity growth have featured the substitution of agricultural chemicals (such as fertilizers and pesticides), energy, seed, and other purchased inputs for labor (farmers' own and hired) and land. Use of agricultural chemicals tripled from 1948 to 1996 (Council of Economic Advisers 1999). During that period the use of feed and seed increased by 64%, the use of energy by 46%, and the use of durable equipment by 44%; the use of labor fell by more than two-thirds; and land use remained roughly constant.

Long-term trends in crop yields provide a crude measure of the contribution of chemicals and breeding to productivity growth. The growth in yields over the last 50 years has been striking: among major US agricultural commodities, for example, yields of corn, wheat, and sorghum have roughly tripled, and yields of soybeans, cotton, and rice have roughly doubled (USDA 1953 and 1999a). Not all long-term growth in yield is attributable to chemicals and breeding; regional shifts in cropping pat-

terns, the introduction of irrigation, improvements in agronomic practices, and other factors have also helped to increase yields. Nevertheless,
chemicals and breeding are important contributors.

The impact of genetic improvements on crop yield has been investigated for several major crops. It has been estimated that the introduction
of and subsequent genetic improvements in hybrid varieties account for
about 80% of the increase in corn yields from 1930 to 1980 and more than
half the increase in soybean yields since 1920 (Duvick 1984; Specht and
Williams 1984; Huffman and Evenson 1993). Genetic improvements have
been estimated to account for half the increase in wheat yields from 1954
to 1979, almost two-fifths of the increase in sorghum yields from 1950 to
1980 (Miller and Kebede 1984; Schmidt 1984; Feyerherm et al. 1988), and
20-35% of the increases in flue-cured tobacco yields from 1954 to 1987
(Babcock and Foster 1991). It has been estimated that, by the middle
1950s, investment in improved corn and soybean hybrids had generated
respective rates of return of around 35-40% and 20% respectively
(Griliches 1958). The rate of return to public potato breeding during the
period 1967-1990 has been estimated to be as high as 80% (Araji et al.1995).

The contribution of pesticides (including herbicides) to agricultural
productivity is less well documented. The methods used in the most
widely known studies (Pimentel et al.1991; Knutson et al.1990) may over-
estimate pest damage. They assume that farmers have little ability to
substitute land, labor, machinery, and other inputs for pesticides and
little flexibility in cropping decisions. Moreover, they tend to rely on data
from experimental plots or expert opinion to estimate crop losses; both
tend to exaggerate differences in yields between use and nonuse of pesticides. The damage-control model of Lichtenberg and Zilberman (1986)
provides a method for inferring crop damage from observed input and
output use. Econometric studies that used that approach to investigate
the aggregate US agricultural economy on the basis of time-series data
found that the use of pesticides roughly halved proportional crop losses,
from about 20% in 1950 to 10% in 1989 (Chambers and Lichtenberg 1994
and 1995). An econometric study of the aggregate US agricultural
economy applied the approach to state-level cross-section data for a single
year and obtained very similar results (Carrasco-Tauber and Moffitt 1992).

A.3   PLANT BREEDING RESEARCH AND DEVELOPMENT IN THE UNITED STATES

R&D in both the public and private sectors have contributed to the
new agricultural technologies that have made productivity growth possible. In 1992, US agricultural R&D expenditures were approximately $6
billion, of which $3.3 billion (56%) came from the private sector, $2.0

billion (33%) came from state agricultural experiment stations (Economic Research Service 1995), and $0.7 billion (11%) came from the USDA.

Broadly speaking, the private sector has concentrated on technologies for which markets provide means of recouping R&D costs, as when patent protection allows private firms to appropriate a share of the benefits generated by new technologies (Huffman and Evenson 1993). In 1992, agricultural chemicals accounted for about one-third of private agricultural R&D, and plant breeding accounted for about 10% (Fuglie et al. 1996). The public sector, in contrast, engages in R&D on technologies for which markets do not provide a viable means of recovering R&D costs. Examples of the former include technologies for which patents would not be enforceable if issued, basic research, and technologies for markets that are simply to small to generate enough revenue to make the technologies sufficiently profitable for the private sector (Huffman and Evenson 1993). Plant-production systems accounted for 35% of state agricultural experiment station spending in 1992 (Fuglie et al. 1996). In 1994, public-sector expenditures on plant breeding amounted to $213 million, about half the $400 million spent by the private sector (Frey 1996; Fuglie et al. 1996).

In pest management, the private sector has undertaken the bulk of R&D for pesticidal substances that can be sold under patent protection. Development of integrated pest management (IPM), which essentially consists of sets of farming practices (such as crop rotation, scouting, field sanitation, tillage methods) that enhance pest control or limit pest damage has been left principally to the public sector. IPM practices are easily imitated, so patents would be unenforceable. Moreover, many IPM programs use combinations of relatively familiar practices and thus might not be considered sufficiently novel to be awarded patent protection.

The relative shares of public and private plant breeding depend largely on the actual or potential size of the market for seeds, which in turn depends on biological and economic factors that influence the feasibility and cost of replicating the performance of superior varieties with saved seed. Important determinants of seed market size include the following:

- The feasibility of producing hybrids with yields sufficiently greater than those of pure varieties, as has been the case with corn and sorghum (but not wheat or soybeans).
- The cash costs of saving seed, including cleaning to eliminate weed seed, storage, and treating to prevent disease and insect damage.
- The implicit costs of saving seed, including the value of forgone harvests of crops not normally harvested for seed (such as forages) and of losses due to delays in replanting.
- The rate at which varieties become obsolete because of the intro-

duction of new varieties that have higher yields, better product quality, and better agronomic traits.

Among major crops in the United States, purchased seed is used for virtually all the corn and sorghum grown and most of the potatoes, cotton, and soybeans (table A.1). Small-grain growers use substantial amounts of saved seed. Saved seed is especially prevalent in wheat. High dockage rates for weed seed (which make careful cleaning profitable) and standard storage practices makes the additional costs of saving seed quite low. In cases where plants are final consumption goods—such

TABLE A.1   Public and Private Sector Breeding Effort, 1994 (PhD-Equivalent Scientist-Years)

Crop	SAES[a]	ARS[b]	Private	Total	Area Planted to Purchased Seed (%)
Corn	27.1	8.2	509.75	545.05	100
Wheat	64.5	11.95	53.95	130.4	20-32
Rice	13.8	6.3	21.9	42	85
Barley	16.4	2.1	13.9	32.4	50
Oats	10.1	2.7	4.9	17.7	40
Sorghum	11.8	2.5	40.8	55.1	95
Other Grains	11.65	0.5	57.75	69.9	
Cotton	19.15	11.65	103.45	134.25	66
Alfalfa	15.2	11.85	41	68.05	97
Other legume forage	9.1	7	2.15	18.25	95
Forage grasses	13.5	14	35.95	63.45	95
Soybean	45	9.6	101.35	155.95	76
Peanut	14	2.5	3.15	19.65	70
Sunflower	0.6	2.56	31.45	34.61	95
Flax	1.3	0	0	1.3	90
Canola	5.7	1	28	34.7	
Other Oilseeds	2.6	0	10.95	13.55	
Potatoes	31	10	9	50	73
Other vegetables	91	16.4	283.65	391.05	85
Sugar	4	15	25	44	
Ornamentals	18	5	64	87	100
Lawn and Turf	15	0	41	56	95
Totals	529	177	1,499	2,205	

[a]State Agricultural Experiment Station

[b]USDA Agricultural Research Service

Source: Breeding effort from Frey (1996). Market shares of corn, soybean, cotton, potatoes, and wheat from Economic Research Service (1997). Market shares of remaining crops from McMullen (1987).

as home and garden uses, golf courses, and other landscaping—virtually all seed is purchased.

Private-sector plant-breeding R&D has been growing rapidly. In nominal terms, private-sector spending on plant breeding rose from $6 million in 1960 to $400 million in 1992 (Economic Research Service 1995). In real terms, private-sector spending increased by a factor of about 13 over this period (an average annual growth rate of 8.3%).

The public and private sectors also differ substantially in the types of breeding R&D undertaken. The public sector concentrates primarily on basic breeding R&D, notably basic research on breeding methods and germplasm enhancement. Each of those general categories accounts for about 30% of public-sector breeding effort, but only 10% of private-sector breeding effort (Frey 1996). The private sector concentrates primarily on cultivar development, that is, preparation of varieties for commercial release.

R&D on transgenic plants exhibits similar differences between the public and private sectors. Most studies have used the number of field trials of transgenic plants as an indicator of R&D effort (Huttner et al. 1995; Ollinger and Pope 1995).

From 1987 to the end of May 1999, the USDA Animal and Plant Health Inspection Service (APHIS) approved 6,531 applications for field trials of transgenic agronomic crops. Data provided by APHIS list the crops involved in 6,522 of them and the types of traits in 6,516. Field trials conducted by private industry focused on herbicide and insect resistance, both of which complement existing product lines of the agrichemical companies responsible for the overwhelming majority (81%) of the trials (table A.3). Universities and nonprofit research institutes focused relatively more effort to basic research (for example, on marker genes) and traits like viral resistance and bacterial resistance, for which pesticidal chemicals are not marketed. Private-sector field trials focused overwhelmingly on corn, which accounted for almost half the industry total (table A.3). Four other major crops—soybeans, cotton, potatoes, and tomatoes—accounted for virtually all of the remainder. The public-sector effort was distributed somewhat more evenly across crops.

A.4 AGRICHEMICAL AND SEED MARKETS IN THE UNITED STATES

USDA estimates that in 1997 US farmers spent $6.7 billion on seed and $8.8 billion on pesticides for agronomic crops alone (Economic Research Service 1997). Seed and pesticide sales have been increasing during the 1990s (figure A.1). Corn, soybeans, wheat, cotton, and potatoes comprise the largest farm-sector markets for seed (table A.2). Corn, soy-

TABLE A.2 US Farm-Sector Sales of Pesticides and Seed for Major Crops, 1997

	Chemical Expenditures per Acre, $	Seed Expenditures per Acre, $	Crop Area, millions of acres	Total Chemical Expenditures, millions of $	Share of Total Chemical Sales, %	Total Seed Expenditures, millions of $	Share of Total Seed Sales, %
Corn	26.87	28.71	80.2	2,155.70	24.4	2,303.32	34.3
Soybeans	28.21	19.66	70.6	1,991.63	22.6	1,388.00	20.7
Wheat	6.32	9.02	71.0	448.65	5.1	640.32	9.5
Cotton	59.47	16.80	13.6	808.79	9.2	228.48	3.4
Rice	68.32	24.15	3.1	208.79	2.4	73.80	1.1
Sorghum	11.71	6.57	10.1	118.27	1.3	66.36	1.0
Barley	9.81	8.96	6.9	67.69	0.8	61.82	0.9
Oats	1.83	9.11	5.2	9.52	0.1	47.37	0.7
Sugar beets	74.15	43.63	1.5	111.23	1.3	65.45	1.0
Peanuts	98.75	74.18	1.4	138.25	1.6	103.85	1.5
Fall potatoes	217.69	156.43	1.4	304.77	1.8	219.00	3.3
Total				*8,827*		*6,711*	

Source: Per acre expenditures by crop from Economic Research Service (1999b). Crop Area from USDA (1999a). Total chemical aand seed expenditures from Economic Research Service (1999b).

TABLE A.3   Gene Function and Crops Involved in Transgenic Field Trials, 1987-May, 1999

Gene Function (Agronomic Crops Only)	Public		Private	
	Number	Percent	Number	Percent
Agronomic properties	39	5	374	6
Herbicide tolerance	55	8	2203	38
Insect resistance	84	12	1838	32
Virus resistance	184	25	447	8
Fungal resistance	61	8	272	5
Product quality	145	20	1434	22
Marker gene	75	10	135	2
Nematode resistance	9	1	4	< 1
Bacteria resistance	53	7	13	< 1
Other	55	8	142	2
Total reporting trait	*725*		*5791*	
Crop (All Plants)				
Corn	81	11	2708	47
Cotton	11	2	488	8
Potato	178	25	539	9
Rapeseed	18	2	216	4
Rice	22	3	73	1
Soybeans	15	2	683	12
Tobacco	162	22	92	2
Tomato	85	12	541	9
Wheat	19	3	61	1
Other	135	19	395	7
Total reporting crop	*726*		*5796*	

Source: USDA (1999c). APHIS data from 1987 to May, 1999.

beans, cotton, and wheat comprise the largest farm sector markets for pesticides, accounting for approximately 60% of total pesticide expenditures (table A.2).

A recent wave of mergers and acquisitions in the seed and agrichemical industries has engendered concern about increasing concentration and its potential impacts on the seed industry and on agricultural R&D more broadly. A number of large agrichemical firms have merged or are merging (table A.4). In addition, the major agrichemical firms have been purchasing agricultural biotechnology and seed firms. Agricultural biotechnology appears to be the principal motivation for mergers and acquisitions in the latter category. Agrichemical firms have several distinct incentives for integrating vertically into the seed industry.

First, genetic engineering creates economies of scale and scope in breeding new varieties. Once identified, single genes can be introduced

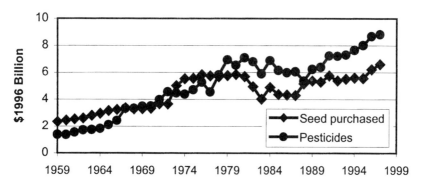

FIGURE A.1   Trends in real (constant dollar) seed and pesticide sales, 1959-1997
Source: Economic Research Service (1999b).

into a number of crops and crop varieties to obtain desired common characteristics (such as the resistance to lepidopteran insects provided by genes that express *Bacillus thuringiensis* (Bt) toxins). Discovering novel genes can cost millions, while inserting the genes into germplasm can cost $10,000-200,000 per gene, depending on the crop.[2] Once a gene has been introduced into a crop line, however, it can be bred into many varieties at no additional cost beyond the normal cost of breeding a new variety. Companies offering a larger number of varieties of a single crop can thus reap economies of scale from a line that contains a gene of interest. Possession of a novel gene, and the specialized knowledge obtained from developing it and breeding it into crops, can lower the cost of inserting the gene into new crops and breeding varieties from the original germplasm. Thus, companies that sell seeds for a large number of crops can reap economies of scope. Furthermore, plant breeding offers new product areas in which to apply proprietary genetic-engineering methods and general expertise derived from pharmaceutical development (that is, economies of scope in the use of human capital and equipment).

Second, genetic engineering may also allow agrichemical firms to

---

[2]Information for estimating the costs of breeding new varieties was obtained from conversations with the following people and their cooperation is greatly appreciated. However, the author is solely responsible for the final estimates.

Carlos Quiros, University of California at Davis (celery)
Walt Fehr, Iowa State University, James Orf, University of Minnesota and
Bill Kenworthy, University of Maryland (soybeans)
Deon Stuthman and Don Rasmussen, University of Minnesota (small grains)
Stephen Baenziger, University of Nebraska (wheat)
Harry Swarz, University of Maryland (small fruits).

TABLE A.4 Mergers and Acquisitions in Agricultural Chemicals, Biotechnology, Seed, and Food or Feed

	Agricultural Chemicals	Biotechnology	Seed	Food or Feed
Monsanto		• Agracetus (1995) • Calgene (1996) • Ecogen (13%) • Millenium Pharmaceutical (joint venture for crop genes)	• DeKalb (1996) • Asgrow (1997) (corn and soybeans) • Holden's Foundation Seeds (1997) • Delta and Pineland (terminated) • Cargill International Seeds, Plant Breeding International (1998)	• Cargill (1998) (joint venture, feed and food. • Nutrasweet Brand
AgrEvo (Aventis)	• Hoechst & Schering (1994) • Hoechst and Rhone-Poulenc (1999) merge to create Aventis	• Plant Genetic Systems (1997) • Plant Tec	• Nunhems, Vanderhave, Plant Genetic Systems, Pioneer Vegetable Genetics, Sunseeds (1997) • Cargill US Seeds (1998)	
Rhone-Poulenc (Aventis)	• Hoechst and Rhone-Poulenc (1999) merger to create Aventis		• Alliance with Limagrain, which owns Nickersons, Vilmorin, Ferry Morse, and others	
Novartis	• Ciba-Geigy and Sandoz merger (1996) • Buys Merck pesticide business (1997)		• Merger brings together Northrup-King, S&G Seeds, Hilleshog, Ciba Seeds • Rogers Seed Co.	• Gerber Foods

Company			
Dow Chemicals	• Dow purchases Eli Lilly's 40% share of DowElanco (1997)	• Mycogen (1996) • Ribozyme Pharmaceuticals, Inc.	• Mycogen buys Agrigenetics (1992) • United Agriseeds becomes part of Mycogen (1996)
Zeneca/Astra		• Mogen International N.V. (1997) • Alliance with Japan Tobacco on rice (1999).	• Advanta (merger of Zeneca Seeds and Vanderhave)
DuPont		• Alliance with Human Genome Sciences (1996) • Curagen (1997)	• Pioneer (1997, 1999) • Hybrinova (1999)	• Optimum Quality Grain (joint venture with Pioneer) • Protein Technologies • Cereal Innovation Centre UK
Empresas La Moderna/Seminis Monsanto in 1997	• DNA Plant Technology (1996)	• Asgrow (1994) (sells corn and soybeans to) • Petoseed (1994) • Royal Sluis • Seminis		• Bionova

Source: Brennan et al. (1999).

augment their product lines in novel ways. For example, Monsanto and AgrEvo have pursued development of herbicide-tolerant crop varieties as a means of enhancing sales of their chemical herbicides.

Third, agrichemical companies' interest in transgenic crops might have been spurred by deceleration in the introduction of new chemical pesticides. The number of new chemical pesticide products registered provides a rough measure of R&D productivity in the years preceding registration. If the average number of products registered per active ingredient remains constant over time, then the number of new products registered will be proportional to the number of new active ingredients introduced into the market. The numbers of new chemical herbicide, insecticide, and fungicide formulated products registered have been lower on average in recent years than in the past (figure A.2). The average annual number of new formulated products of each type introduced in 1990-1998 was about half the average during each of the 2 preceding decades.

Fourth, regulation of chemical pesticides under FIFRA and public controversies over the use of pesticides may have made investment in transgenic crops more attractive. In 1993, EPA estimated that meeting data requirements for registering a chemical pesticide cost $10.6 million; that corresponds to a cost of $11.7 million in 1998 dollars. Meeting data requirements for transgenic pest-protected plants, by contrast, has been estimated to cost between $0.07-1.17 million in 1998 dollars, depending on the source of the pesticidal substance, the presence of wild relatives in the United States, and the extent of information available on the characteristics and function of the gene(s) introduced (EPA 1994d). Currently commercialized transgenic pest-protected plants feature more efficient targeting of pests than chemical pesticides and thus have the potential for less-extensive offsite and nontarget impacts.

The wave of consolidation has raised two kinds of economic concerns about potential loss of competition. First, increased concentration might allow firms in the agricultural-supply industry to exert market power, reducing farm income and increasing food prices. Second, increased concentration might lead firms to reduce R&D and thus dampen growth in agricultural productivity. The principal incentive for reducing R&D is to protect sales of existing products. Just and Hueth (1997), however, have noted that the opposite incentive holds when new products are complements of existing ones. In that case, new products increase demand for existing ones, so that introducing them leads to increased profit. As a result, firms with market power might engage more heavily in R&D than firms in a more competitive market would. Herbicide-tolerant crops are a good example of such a new product. Such crop varieties as Roundup Ready or Liberty Link corn or soybeans allow farmers to substitute

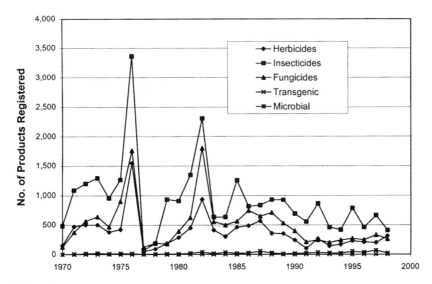

FIGURE A.2   New Formulated Pesticide Products Registered, 1970-1998.
Source: EPA 1999d.

Monsanto's or AgrEvo's herbicides for those sold by their competitors, increasing Monsanto's and AgrEvo's overall profit.

While concentration appears to be increasing in a number of major seed markets (see below), it is not clear whether this increase in concentration is attributable to the wave of recent mergers. The seed industry underwent a similar wave in the late 1970s, as petrochemical companies (Shell and Arco), pharmaceutical manufacturers (Ciba-Geigy, FMC, Pfizer, Sandoz, and Upjohn), and other chemical firms (W.R. Grace) acquired both biotechnology firms and seed companies (McMullen 1987; Kloppenburg 1988; Hayenga and Kimle 1992). Agricultural biotechnology and the prospect of expanding markets for US farm exports were principal motivations. The fall in farm exports and the ensuing farm financial crisis of the 1980s led petrochemical and specialty-chemical firms to leave this industry and to sell their biotechnology and seed subsidiaries to agrichemical companies. Some pharmaceutical firms also left the agrichemical business to concentrate on human-health and veterinary products. As a result, some of the current wave of mergers consist of changes in parent companies rather than new consolidation.

Two measures are widely used to determine the degree of concentration in an industry. The four-firm concentration ratio (C4) is the sum of the shares of total sales in an industry accounted for by the four largest firms in the industry. The Herfindahl-Hirschman Index (HHI) is the sum of the squared percentages of the industry's total sales accounted for by the individual firms in the industry. If the industry is completely monopolized, the HHI is at its maximum, 10,000. If there are N firms in the industry, each with an equal share of sales, the HHI equals 10,000/N. As the number of firms in the industry increases, the HHI falls toward zero. The HHI is the preferred measure of concentration because under some conditions it is proportional to the markup of price over marginal cost and so indicates the excess profit due to the exercise of market power (Cowling and Waterson 1976). The US Department of Justice considers an HHI over 1,800 to indicate market concentration worthy of consideration for potential antitrust action.

Published US data are available on four major seed markets—corn, soybeans, cotton, and vegetables. Published estimates of corn seed market shares of the 6-10 leading firms are available for about half the years since 1973—before the first merger wave that affected the US seed industry (table A.5). The market has become more concentrated, but the increase in concentration, is due almost entirely to increases in the market share of a single firm, Pioneer Hi-Bred (acquired in 1999 by DuPont), rather than to mergers and acquisitions. Published estimates of soybean-seed market shares are available for 1988 and 1997 (table A.6). Taking into account the market share of farmer-saved seed, it appears that the most recent mergers and acquisitions have increased concentration in this market, although the degree of concentration is still not high. The US cotton-seed market is highly concentrated because a single firm, Delta and Pine Land, controls 70-75% of total sales (Hayenga 1998). Data on market shares of the top two firms alone reported by Hayenga (1998) imply an HHI of at least 5,300, which indicates a high degree of concentration. As with corn, however, the concentration predated Monsanto's entry into the industry and remains unaffected by the termination of the proposed Monsanto-Delta and Pine Land merger. The vegetable seed-market also appears highly concentrated. The Mexican conglomerate Empresas La Moderna (ELM) accounts for about 40% of US vegetable-seed sales, and mergers and acquisitions have given the agrichemical firm Novartis a market share about half that of ELM (Friedland and Kilman 1999). Both companies have built their market share primarily through mergers and acquisitions. In this case, mergers and acquisitions do appear to have resulted in increased concentration.

Estimates of pesticide market shares were not publicly available. However, markets for some specific pesticides tend to exhibit substantial concentration. For example, Hayenga (1998) estimates that in 1998 the top four firms accounted for about 80% each of sales for soybean, corn, and cotton herbicides.

Publicly available data are insufficient to determine whether all possible seed markets exhibit concentration. The fragmentary data that are available do not clearly indicate that the current merger wave has re-

TABLE A.5 Concentration in the US Corn-Seed Market, 1973-1997

Market Shares[a]

Company	1973	1975	1977	1978	1979	1980	1981	1982	1983	1988	1996	1997
Pioneer	23.8%	24.6%	30.9%	26.2%	32.9%	36.9%	35.0%	38.8%	38.1%	35.0%	41.0%	42.0%
Monsanto												14.0%
DeKalb	21.0%	18.8%	15.8%	17.9%	13.3%	13.0%	13.9%	12.2%	10.3%	9.0%	10.1%	
Asgrow										3.1%	2.0%	
Trojan	5.9%	6.8%	4.2%	5.4%	3.8%	2.0%						
Novartis												9.0%
Northrup-King	6.1%	4.7%	3.8%	3.3%	3.8%	4.9%	2.8%	2.6%	2.5%	4.5%	5.0%	
Funk	8.8%	8.9%	6.4%	8.1%	6.7%	5.7%	5.8%	5.2%	3.9%	3.4%		
Ciba											3.1%	
Zeneca/ICI										3.4%	2.9%	
Aventis												7.0%
Cargill	4.8%	3.9%	4.1%	4.6%	3.3%	4.7%	5.8%	5.4%	4.2%	3.9%	3.3%	
Dow/Mycogen										1.4%	4.3%	4.0%
Jacques/Agrigenetics		1.7%	1.9%	2.1%	2.7%	2.2%				1.8%		
Golden Harvest		1.8%	2.5%	3.1%	2.9%	1.3%	2.8%	2.3%	2.6%	2.5%	2.3%	4.0%
Other	29.6%	29.8%	30.4%	29.3%	30.6%	29.3%	33.9%	33.6%	38.4%	32.0%	25.6%	20.0%
C4	59.7%	59.1%	57.3%	57.6%	56.7%	60.5%	60.5%	61.6%	56.5%	52.4%	60.4%	72.0%
HHI	1,180	1,127	1,304	1,148	1,360	1,620	1,501	1,723	1,604	1,386	1,865	2,122

[a]Except C4 and HHI.

Source: Data for 1973-1983 cited in McMullen (1987). Data for 1988 from Hayenga and Kimle (1992). Data for 1996 from Kalaitzandonakes (1997). Data for 1997 from Hayenga (1998).

TABLE A.6 Concentration in the US Soybean-Seed Market

Company	Market Share[a]	
	1988	1997
Monsanto	2.6%	14.4%
DeKalb	4.2%	0.0%
Asgrow	11.3%	0.0%
Stoneville		0.0%
Pioneer Hi-Bred	10.4%	14.4%
Novartis	5.8%	3.8%
NC		0.0%
Dow/Mycogen		3.0%
FS	1.7%	
Stine	2.6%	3.0%
Jacques	1.3%	0.0%
Other brands	12.9%	29.6%
Public brands	23.2%	7.6%
Saved seed	24.0%	24.0%
Total	*100.0%*	*100.0%*
C4	*31.7%*	*35.7%*
HHI	*1010*	*1386*

[a]Except C4 and HHI.

Source: Data on 1988 from Hayenga and Kimle (1992). Data on 1997 from Hayenga (1998).

sulted in a significant increase in concentration in major seed markets—such as those for corn, soybean, and cotton in the United States—or the overall world seed market. It remains possible that concentration has increased significantly in specific submarkets, such as corn seed in a specific growing region, or in markets for crops with smaller planted area. Moreover, firms might be able to exercise market power because of their ownership of key inputs into seed production, such as gene-insertion techniques, genes for herbicide tolerance or proteins that confer pest resistance, or germplasm from widely used inbred lines.

Similarly, it is not clear whether vertical integration of seed companies into agrichemical companies has increased concentration in the crop-protection market. For example, the introduction of herbicide-tolerant crops has increased Monsanto's share of the markets for herbicides on corn, soybeans, and cotton while reducing the share of the market leaders, thereby reducing the degree of concentration in these markets. In 1996, American Cyanamid and DuPoint accounted for an estimated 60% and 20%, respectively, of soybean-herbicide sales. The availability of Roundup Ready soybeans has cut those market shares in half and led both companies to cut herbicide prices (Kilman 1999). Similarly, the introduction of Bt corn and cotton has increased competition in the markets for corn and cotton insecticides and led to insecticide-price cuts.

By the same token, competition from chemical pesticides limits the extent to which firms can exercise market power in the sales of transgenic pest-protected plants. Adoption of Bt cotton, for example, has been limited in areas where insecticide costs have traditionally been low, notably where insect pressure has traditionally been low. As a result, Monsanto's sales of and profit from Bt cotton have been high mainly in Alabama, Mississippi, Georgia, and Florida (Falck-Zepeda et al. 1999).

There appear to be relatively few firms (less than 50) presently engaged in R&D on transgenic crop protection (including microbials and biological controls in addition to transgenic pest-protected crops). *Genetic Engineering News* (1998) estimated that in 1997 there were 492 companies worldwide engaged in agricultural biotechnology and 186 in pesticide biotechnology. Roughly 45-50 were engaged in breeding plants with pesticidal, growth-regulator, or other traits regulated under FIFRA; nine of these were large multinationals (inclusive of subsidiaries). The National Biological Impact Assessment Program housed at the Virginia Polytechnic Institute lists about 45 US companies in engaged in pesticide biotechnology R&D, of which 11 are major multinationals (inclusive of subsidiaries), and about half the remainder appear to be small startup companies that have fewer than 20 employees.

There is some evidence that the recent wave of mergers and acquisitions has increased concentration in transgenic-plant R&D, at least at the late pre-commercialization stages measured by field trial activity. In 1988-1998, the four leading firms accounted for 63-87% of field trials approved each year. In 1998, the four leading firms—Monsanto, AgrEvo, Pioneer Hi-Bred, and DuPont—accounted for 79% of approved field trials. Mergers and acquisitions in the industry raised the HHI for field trials in 1998 from 1,608 to 2,182 (Brennan et al. 1999).

A.5 COSTS OF REGULATING TRANSGENIC PEST-PROTECTED PLANTS

All transgenic pest-protected plant products will be subject to some level of regulatory oversight prior to commercialization, regardless of whether EPA's proposed rule regulating them as "plant pesticides" is implemented. In most instances, commercialization requires clearance from USDA in the form of a determination of nonregulated status under the Federal Plant Pest Act. Foods derived from transgenic plants, including those with novel elements such as compounds with pesticidal or growth-regulator activity, are subject to review by FDA through its voluntary consultation process. It should be noted, however, that FDA has the authority to require formal pre-market review in any case where it is deemed necessary by the agency. This review could take the form of a food additive petition or a GRAS clearance process, both of which are designed to demonstrate the safety of any added substances (see chapters 1 and 4).

A.5.1 Regulatory Costs

Regulation of pesticidal substances in transgenic pest-protected plants as plant-pesticides by EPA will probably involve more extensive provision of data prior to commercialization and will therefore be more expensive than regulation by APHIS or oversight under FDA's consultation process. There are two reasons for the likely extra cost:

- EPA will review pesticidal substances in transgenic plants for potential health and environmental effects not considered by USDA-APHIS or FDA, including acute and chronic toxicity to nontarget organisms (both vertebrate and invertebrate), potential for water pollution, and similar types of environmental effects.
- EPA tends to rely more heavily on test data than USDA-APHIS and tends to require more extensive premarket submission of data than that submitted to FDA under the consultation process.

In 1993, EPA estimated that the costs of testing plant-pesticides would be around $64,000-1,070,000, depending on the origin of the pesticidal substance, the presence of wild relatives, and the extent of available information on the characteristics and function of the gene(s) involved (EPA 1994d). Those estimates are based on experience with such products as Bt and viral coat proteins and are therefore likely to understate the costs of testing new generations of products. Current transgenic pest-protected plants do not create exposures of a qualitatively new type: viral coat protein is present naturally in most plants, and Bt has been used in microbial form for a long time and is a familiar product. New generations of products, in contrast, will probably use less familiar proteins, so one would expect EPA to require more extensive testing of future transgenic pest-protected plant products.

EPA's estimates of testing costs also underestimate the costs of regulatory compliance because they ignore the cash costs and implicit costs of management time needed for overseeing the regulatory process and interacting with EPA staff. Those costs are likely to be higher for smaller entities, such as biotechnology startup companies, small to medium seed companies, and public-sector breeders. Major agrichemical firms have staff dedicated to regulatory affairs who have extensive familiarity with the regulatory process and EPA staff; smaller entities do not. The incremental cost of complying with regulation of a single new product is thus considerably less for major agrichemical companies than for small entities.

EPA's published estimates of testing costs (EPA 1994d) were used to estimate the costs of meeting potential testing requirements imposed by USDA-APHIS regulation, safety reviews and data submissions under the FDA consultation process, as well as testing under pesticide regulation (table A.7). EPA's figures were converted to 1998 dollars by using the implicit GDP price deflator reported by the Council of Economic Advisers (1999). EPA estimates of Tier II and III testing costs for microbial pesticides were used to estimate potential costs of further testing new, unfamiliar products for human health effects and mammalian toxicity. EPA's estimated costs of providing material for testing for biological fate were then added.

Baseline data likely to be submitted to USDA-APHIS and FDA as well as EPA consists of product analysis (including crop residue), Tier I biological fate, acute oral toxicity, and digestibility for a total of about $20,000. Additional costs of regulating pesticidal substances in transgenic pest-protected plants as pesticides beyond that required for Bt and viral coat proteins would include the following:

- Testing for effects on nontarget organisms. The cost of full batteries of tests ranged from about $76,000 for Tier I tests to over $410,000 for Tier I-III tests.
- More extensive testing on biological fate. The cost of full batteries of tests ranged from about $46,000 for Tier II tests to over $735,000 for Tier II and III tests.
- More extensive testing on human health and mammalian toxicity. The costs of testing ranged from about $10,000 for hypersensitivity and specific allergen screening to $1,667,000 for a full battery of testing, including Tier II and III toxicity testing.

Overall, then, according to EPA cost estimates the additional testing costs involved in regulating pesticidal substances in transgenic pest-protected plants as plant-pesticides (that is, the testing required by EPA beyond the data submitted to USDA-APHIS and FDA) could total as much as $2.8 million. As noted above, that total is modest compared with that for testing chemical pesticides. It might nevertheless be substantial relative to the cost of breeding new varieties and thus influence both the types of transgenic crops developed and entry by less well-capitalized entities. Moreover, EPA's estimates may understate the actual costs of conducting the required testing: Unpublished Monsanto estimates, for example, indi-

TABLE A.7   Estimated Costs of Regulatory Testing for Transgenic Pest-Protected Plants in 1998 Dollars

	EPA Estimated Testing Costs	Increment in Breakeven Expected Annual Sales
Baseline data requirements:		
Product analysis, including crop residues	$10,952	$2,417
Biological fate (basic)	$1,923	$424
Acute oral toxicity and digestibility	$7,240	$1,598
Total	*$20,115*	*$4,439*
Biological fate:		
Tier II[a]	$45,647	$10,073
Tier III[b]	$689,650	$152,182
Total	*$735,297*	*$162,255*
Human health and mammalian toxicology:		
Hypersensitivity and specific-allergen testing	$10,138	$2,237
Tier II (microbial)[c]	$276,948	$61,113
Tier III (microbial)[d]	$1,379,795	$304,474
Total	*$1,666,881*	*$367,824*
Nontarget organisms:		
Tier I[e]	$76,123	$16,798
Tier II[f]	$86,985	$19,195
Tier III[g]	$248,187	$54,767
Total	*$411,295*	*$90,759*

[a]tests for hybrid and pollen viability

[b]tests for selective advantage (host range and growth and development) and dispersion (including field testing)

[c]tests for acute and subchronic toxicity/pathogenicity

[d]tests for reproductive effects, oncogenicity, immunodeficiency, and primate infectivity/pathogenicity

[e]tests for honey bee toxicity, avian oral LD$_{50}$, avian dietary LC$_{50}$, and nontarget insect effects.

[f]tests for freshwater fish LC$_{50}$, acute freshwater organisms EC$_{50}$, and acute estuarine and marine organism LC$_{50}$.

[g]tests for avian reproduction

cate that the costs of providing data to meet regulatory requirements for Bt corn amounted to nearly $3.8 million in addition to 21.5 person-years of staff time.

If pesticidal substances in transgenic plants were not considered pesticides for purposes of FIFRA and the FFDCA, it is possible that the costs associated with the FDA process would increase. In particular, for at least some of these substances, FDA might well require submission of a formal food additive petition or data equivalent to that required for such a petition in order to demonstrate that any added substances were GRAS. The costs associated with such a submission would depend on the amount and type of data required to demonstrate safety at the anticipated level of exposure. In 1997, FDA estimated the costs of data requirements for food additive petitions using surveys from four food companies (FDA 1997b). Adjusted to 1998 dollars using implicit GDP price deflator reported by the Council of Economic Advisors (1999), these costs ranged from $174,787 to $1,359,456. Comparing these costs to the costs estimated in table A.7 for EPA data requirements under FIFRA and FFDCA, the additional cost associated with the EPA process would be in the range of $1.5 to $2.6 million. In the event FDA were to require an environmental assessment (EA) under NEPA in conjunction with the food additive petition, the costs of developing data to support the EA might fall in the range of $2500 to $50,000 (FDA 1997a), which would reduce the additional cost of the EPA process by the same amount

A.5.2   Comparison of Regulatory Costs to Costs of Breeding a New Variety

To assess the extent to which regulation might discourage R&D related to crops that would have small potential markets and R&D by smaller entities, I compare the potential costs of meeting regulatory requirements with the fixed costs of breeding a new variety. I ignore the variable costs of producing seed for commercial sale, which depend on the size of the market. I consider only the costs of developing a crop to the point where it would be ready to scale up production for commercial sale. The costs of regulatory compliance are a form of fixed cost in that regulatory approval is needed before commercialization. It is thus appropriate to compare regulatory compliance costs with the fixed costs of breeding a new variety.

The costs of crop breeding depend on the costs of running a breeding operation, the time required to develop a new variety sufficiently for market introduction, and the success rate of new varieties. A simple model can be used to indicate how those factors influence the cost of developing a new variety. Let C denote the cost of running a breeding facility for a year, including direct costs and overhead; for simplicity, C is assumed to be constant (in real terms). Let T be the expected time required to develop a variety to the point where it is marketable. Let N denote the average number of years between successful introductions of new varieties, so that an average of $1/N$ new varieties are introduced every year. Let r be the (real) interest rate. The average cost of a new variety can be expressed as $\frac{CN\,[e^{rt}-1]}{rT}$. For example, a breeding operation with a direct annual cost of $100,000 plus an overhead rate of 50%, facing a real interest rate of 4%, needing 10 years to develop a new variety, and introducing a new variety every 3 years would develop new varieties at an average cost of about $553,000 each.

The literature contains few estimates of breeding costs. McMullen (1987) estimated that the average time to breed new varieties with traditional methods ranged from 7.5 years for corn and safflower to a high of 14-15 years for squash and watermelon and almost 19 years for cauliflower (table A.8). He cited published estimates indicating costs of around $1.5-3.0 million for developing a new crop variety, corresponding to $2.0-4.0 million in 1998 dollars. He estimated the direct cost of public breeding programs at around $250,000 ($340,000 in 1998 dollars) per year. The cost of introducing a new tomato variety has been estimated at $315,000-630,000 in 1998 dollars (NRC 1989).

Conversations with public-sector crop breeders indicate that the costs of new varieties can vary substantially. The cost of breeding a new small fruit (such as strawberry or raspberry) variety in a public program, calculating with the model presented above, appears to be about $200,000. Breeding soybeans appears to exhibit economies of scale, that is, the cost per variety is lower in larger programs. Large programs appear to breed new varieties at a cost of $215,000-285,000, small programs at about $550,000. The cost of breeding celery in a public program is about $425,000 per variety. The cost of breeding small grains is considerably higher: on the order of $2.1 million for oats and $2.8-3.0 million for wheat.

The preceding analysis ignores the cost of developing germplasm and considers only the cost of developing a new variety from existing germplasm. It was not possible to estimate the cost of developing new germplasm for use in breeding programs via traditional breeding methods or genetic engineering. That would require estimating the costs of screening germplasm and identifying useful traits and the costs and success rates of introducing identified traits into existing germplasm. None of those dimensions could be estimated for this study, although unpublished Monsanto estimates indicate that the total costs of developing Bt corn were in the range of $10 to $25 million, inclusive of germplasm development (molecular biology, gene expression, transformation) and development of commercial varieties (insect evaluation, event screening, field evaluation, and product development). In general, however, one would expect germplasm with useful traits to be used in developing a large number of varieties. As a result, the average per-variety cost of germplasm is likely to be small. For example, if Monsanto's Bt corn germplasm were used in 1,000 varieties, the average per variety cost of

TABLE A.8 Time Required for Traditional Breeding of New Crop Varieties

Crop	Cross to Date of Determination, years	Date of Determination to Application, years	Total, years
Barley	7.0	3.4	10.4
Bean	8.0	3.3	11.3
Cauliflower	11.0	7.5	18.5
Corn	5.5	2.0	7.5
Cotton	8.0	4.2	12.2
Lettuce	7.0	2.6	9.6
Oats	8.8	2.1	10.9
Onion	9.0	2.9	11.9
Peas	7.0	4.0	11.0
Rice	6.0	2.8	8.8
Safflower	6.0	1.7	7.7
Soybean	6.2	3.0	9.2
Squash	11.0	3.7	14.7
Tobacco	8.5	2.6	11.1
Tomato	8.3	1.4	9.7
Watermelon	8.5	5.0	13.5
Wheat	8.0	2.8	10.8

Source: McMullen (1987).

that germplasm would be under $10-25,000 (once varietal development costs were subtracted from Monsanto's estimate).

Conversations with specialists in the field indicate that, once a novel gene is identified, inserting it into crop germplasm via genetic engineering could cost $10,000-200,000. If it is used for a single variety only, the model presented above indicates that the additional gene could increase the fixed cost of breeding a new variety by $37,000-690,000. Once present in germplasm, however, the gene is available for use in multiple crosses, as is any other germplasm used in breeding. For example, genes for single Bt toxins have already been used in dozens of crop varieties. The gene is likely to be used in multiple varieties, so the increase in the fixed cost of breeding will generally be considerably lower because the cost per variety decreases geometrically with the number of varieties in which the gene is used.

Developing a new variety will be economically viable if the present value of its sales at least covers the cost of development. Thus, expected annual sales of the variety and its expected lifetime in the market will influence R&D decisions. Let S denote expected annual sales of the variety, assumed constant (in real terms) for simplicity, and D denote the expected lifetime in the market. Then the breakeven level of expected annual sales needed to justify development of a variety is $\frac{CN\,erT-1}{T\,1-e^{-rD}}$. As an example, suppose that the varieties of each of the crops mentioned above lasted 5 years in the market (varieties of soybean, oats, and wheat have had average market lifetimes of 3-5 years recently.) At a real interest rate of 4%, breakeven annual sales would be about $45,000 for small fruits, $47,000-120,000 for soybean, $70,000-140,000 for tomatoes, $94,000 for celery, $465,000 for oats, and $620,000-660,000 for wheat.

The model indicates that developing new varieties will be more attractive economically when the development time T is shorter, the success rate is greater (N is smaller), and the variety is expected to last longer on the market (D is greater).

Use of genetic engineering will increase the size of the market needed to break even. The following procedure was used to estimate the increment in breakeven annual sales. If K denotes the per-variety cost of transgene insertion, transgene insertion increases breakeven annual sales needed to justify development of a new variety by $\frac{rK}{1-e^{-rD}}$. The model indicates that this increment in breakeven sales is greater when the transgene-insertion cost, K, is higher and the expected lifetime of the variety on the market D is shorter. If a transgene is used for only a single variety, it will increase breakeven annual sales by $8,000-152,000. A gene will probably be used for more than one new variety, so the actual increment in breakeven annual sales will generally be considerably smaller, as noted above.

The same procedure can be used to estimate the impact of regulatory compliance costs on breakeven sales. In this case, K represents regulatory testing costs, which are assumed to be incurred in a lump sum at the time of product introduction. As before, I assume a product lifetime of 5 years and a real interest rate of 4%. Baseline data likely to be required under any form of regulation would probably increase the breakeven expected annual sales needed to justify R&D investment by only $4,400 (table A.7). EPA's estimates based on viral coat proteins and Bt crops indicate increments in breakeven sales of around $14,000-236,000. Regulation of transgenic pest-protected plant products with novel, unfamiliar genes as pesticides could increase breakeven sales by $620,000 or more (table A.7).

A.6 SUMMARY

Regulating pesticidal substances in transgenic pest-protected plants as pesticides could create substantial barriers to R&D related to minor-use crops and to entry by small entities. Baseline regulatory testing requirements appear to impose relatively low additional fixed costs on the commercialization of new crop varieties. The fixed costs of complying with pesticide regulatory-testing requirements, in contrast, are quite large relative to the fixed costs of developing new crop varieties. For example, the estimates presented here suggest that the sales volume needed to offset the cost of regulatory testing under FIFRA is almost triple the sales volume needed to meet the cost of developing a new variety of wheat from existing germplasm and 14 times the sales volume needed to meet the cost of developing a new small fruit variety from existing germplasm. As a result, regulating transgenic pest-protected plant products as "plant-pesticides" is likely to increase the expected annual sales needed to justify R&D investment substantially, making R&D related to crops with small seed markets less attractive and making it more difficult for smaller, less well-capitalized entities to enter the market.

Appendix B

Example of Data Submitted to Federal Agencies

The committee had the opportunity to review scientific studies that reflect the type of data submitted by Monsanto to EPA in support of registration for Bt Corn, Bollgard Cotton, and NatureMark NewLeaf Potato. The amount of data can be extensive (studies are typically 20 to 150 pages, single spaced, typed, 1" margins, 12 pt font), and as such, the committee did not have the time to thoroughly analyze all of the data provided. Although the specific studies may change depending on the transgenes and the plant species, the following list gives an indication of the type of data typically submitted for some transgenic pest-protected plants containing Bt transgenes.

Examples of Studies Submitted for Transgenic Pest-Protected Plants Containing Bt Transgenes

Corn

Molecular characterization of insect protected corn line MON 810.

Evaluation of insect-protected corn lines in 1994 U.S. field test locations.

Assessment of the equivalence of B.t.k HD-1 protein produced in several insect protected corn lines and *Escherichia coli*.

Compositional comparison of *Bacillus thuringiensis* subsp. *kurstaki* HD-1 protein produced in ECB resistant corn and the commercial microbial product, DIPEL.

Assessment of the equivalence of *Bacillus thuringiensis* subsp. *kurstaki* HD-1 protein produced in *Escherichia coli* and European corn borer resistant corn.

A dietary toxicity study with MON 80187 meal in the northern bobwhite.

Aerobic soil degradation of *Bacillus thuringiensis* var. *kurstaki* HD-1 protein.

Acute oral toxicity study of Btk HD-1 tryptic core protein in albino mice.

Assessment of the in vitro digestive fate of *Bacillus thuringiensis* subsp. *kurstaki* HD-1 protein.

Stability of the Cry1A(b) insecticidal protein of B.t.k. HD-1 in sucrose and honey solutions under non-refrigerated temperature conditions.

Evaluation of the dietary effects of purified B.t.k. endotoxin proteins on honey bee larvae.

Evaluation of the dietary effects of purified B.t.k. endotoxin proteins on honey bee adults.

Activated B.t.k. protein: a dietary toxicity study with green lacewing larvae.

Activated B.t.k. protein: a dietary toxicity study with parasitic hymenoptera (*Brachymeria intermedia*).

Activated B.t.k. protein: a dietary toxicity study with ladybird beetles.

Evaluation of European corn borer resistant corn line MON 801 as a feed ingredient for catfish.

Cry1A(b) insecticidal protein: an acute toxicity study with the earthworm in an artificial soil substrate.

Effects of the *Bacillus thuringiensis* insecticidal proteins Cry1A(b), Cry1A(c), Cry3A on *Folsomia candida* and *Xenylla grisea* (Insecta: Collembola).

Supplemental submission to MRID 43665502 on the expression of the Cry1A(b) protein in insect-protected line MON 810.

Supplemental submission on the tissue expression and corn earworm (*Helicoverpa zea*) efficacy of the Cry1A(b) protein in insect-protected corn.

Chronic exposure of *Folsomia candida* to corn tissue expressing Cry1A(b) protein.

Corn pollen containing the Cry1A(b) protein: a 48-hour static-renewal test with Cladoceran (*Daphnia magna*).

Cotton

Determination of copy number and insert integrity for cotton line 531.

Gene expression and compositional analysis from field-grown insect resistant cotton tissues.

Assessment of equivalence between *E. coli*- produced and cotton-produced B.t.k. HD-73 protein and characterization of the cotton–produced B.t.k. HD-73 protein.

Charcterization of purified B.t.k. HD-73 protein produced in *Escherichia coli*.

Sensitivity of insect species to the purified Cry1Ac insecticdal protein from *Bacillus thuringiensis* var *kurstaki* (B.t.k. HD-73).

Stability of the Cry1Ac insecticidal protein of *Bacillus thuringiensis* var *kurstaki* (B.t.k. HD-73) in sucrose and honey solutions under non-refrigerated temperature conditions.

Evaluation of the dietary effect(s) of purified B.t.k. endotoxin proteins on honey bee larvae.

Evaluation of the dietary effect(s) of purified B.t.k. endotoxin proteins on honey bee adults.

B.t.k. HD-73 protein: A dietary toxicity study with parasitic hymenoptera (*Nasonia vitripennis*).

B.t.k. HD-73 protein: A dietary toxicity study with ladybird beetles (*Hippodamia convergens*).

B.t.k. HD-73 protein: A dietary toxicity study with green lacewing larvae (*Chrysopa carnea*).

A dietary toxicity study with cotton seed meal in the northern bobwhite.

B.t.k. HD-73 protein dose formulation and determination of dose for an acute mouse feeding study MD 92-493.

Acute oral toxicity of *Bacillus thuringiensis* var. *kurstaki* (Cry1Ac) HD-73 protein in albino mice.

Assessment of the in vitro digestive fate of *Bacillus thuringiensis* var *kurstaki* HD-73 protein.

Aerobic soil degradation of *Bacillus thuringiensis* var *kurstaki* HD-73 protein bioactivity.

Potato

Molecular characterization of CPB resistant Russet Burbank Potatoes.

Determination of the expression levels of B.t.t. and NPTII proteins in potato tissues derived from field grown plants.

Equivalence of microbially-produced and plant-produced B.t.t. also called Colorado Potato Beetle active protein from B.t.t.

Characterization of Colorado Potato Beetle active B.t.t protein produced in *E. coli.*

Characterization of major tryptic fragment from Colorado Potato Beetle active protein from B.t.t. .

Compositional comparison of Colorado Potato Beetle active B.t.t. produced in Colorado Potato Beetle resistant potato plants and commercial microbial products.

Sensitivity of selected insect species to the Colorado Potato Beetle active protein from B.t.t.

Stability of Colorado Potato Beetle active B.t.t. protein in sucrose and honey solution under non-refrigerated temperature conditions.

Evaluation of the dietary effect(s) of purified B.t.t. protein on honey bee larvae.

Evaluation of the dietary effect(s) of purified B.t.t. protein on honey bee adults.

B.t.t. protein: a dietary toxicity study with parasitic hymenoptera (*Nassonia vitripennis*).

B.t.t. protein: a dietary toxicity study with ladybird beetles (*Hippodamia convergens*).

B.t.t. protein: a dietary toxicity study with green lacewing larvae.

B.t.t. protein: a dietary toxicity study with Russet Burbank potato in the northern bobwhite.

Colroado potato beetle active B.t.t protein dose formulation, dose confirmation, and dose characterization for albino mice acute toxicity study (ML-92-407)

Acute oral toxicity of B.t.t. protein in albino mice.

Assessment of the metabolic degradation of the Colorado Potato Beetle active protein in simulated mammalian digestive models.

Aerobic soil degradation of Colorado Potato Beetle active protein from B.t.t.

Strategies for Colorado Potato Beetle Resistance Management in NewLeaf Potatoes.

Appendix C

Committee on Genetically Modified Pest-Protected Plants

PUBLIC WORKSHOP

NATIONAL RESEARCH COUNCIL
BOARD ON AGRICULTURE AND NATURAL RESOURCES

May 24, 1999

AGENDA

8:30 – 9:00 am	**Introduction to Workshop** Perry Adkisson, Committee Chair
	Overview of Coordinated Regulatory Framework Stanley Abramson, Committee Member
9:00 – 10:00	**Panel A: Molecular Biologists/Traditional Plant Breeders Panel**

➤ Do transgenic pest-protected plants pose new, or different risks and benefits compared to traditionally bred plants?
➤ What are the differences, if any?
➤ What are the similarities?
➤ How should the regulatory scheme account for the differences or similarities?

241

Marlin Rice
Professor, Department of Entomology
Iowa State University

James Cook
R. James Cook Endowed Chair in Wheat Research,
Department of Plant Pathology, Crops and Soils
Washington State University

Doreen Stabinsky
Assistant Professor
Department of Environmental Studies
California State University

Richard Allison
Professor, Department of Botany and Plant Pathology
Michigan State University

10:00–10:15   Break

10:15–11:15   **Panel B: Commodity and Sustainable Agriculture Panel**

➤ What are the effects of genetically modified pest-protected plants on farming practices and/or farmers?
➤ What are the effects of the regulation of these plants on farming practices and/or farmers?
➤ Do GMPP plants have a role to play in sustainable agriculture for the future?

Robert Mustell
Vice President for Marketing
National Corn Growers Association

Tim Debus
Director for Industry Coalitions
United Fresh Fruit and Vegetable Association

Kathryn DiMatteo
Executive Director
Organic Trade Association

Rick Welsh
Policy Analyst
Wallace Institute

11:15 – 12:15   **Panel C: Ecological Effects Panel**

➤ What evidence is available concerning potential
ecological risks or benefits associated with
GMPP plants? (e.g. non-target species, resistance management, habitat considerations,
biodiversity, reduced pesticide use)
➤ How should the regulatory framework address
these risks and benefits?

Harold Coble
Professor, Department of Crop Science
North Carolina State University

Peter Kareiva
Senior Scientist, Fish and Wildlife Service
Department of Interior

Phil Regal
Professor
Department of Ecology, Evolution, and Behavior
University of Minnesota

Guenther Stotzky
Professor, Department of Biology
New York University

12:15 – 1:15pm   Lunch break

1:15 – 2:15   **Panel D: Animal/Human Health Effects Panel**

➤ What evidence is available concerning potential
risks or benefits that GMPP plants pose to livestock or human health? (e.g. allergenicity, toxicity, reduced pesticide exposure)
➤ How should the regulatory framework address
these risks and benefits?

Steven Druker
Executive Director
Alliance for Biointegrity

Hugh Sampson
Professor, School of Medicine
Immunobiology Center
The Mt. Sinai School of Medicine

John Trumble
Professor, Department of Entomology
University of California, Riverside

Vasilios Frankos
Principal
Environ, Life Sciences Division

2:15 – 3:45   **Panel E: Coordinated Regulatory Framework Perspectives Panel**

➤ How effectively does the coordinated framework for the regulation of biotechnology address the scientific risks?
➤ What are the weaknesses?
➤ What are the strengths?
➤ What role does each agency play in the framework?
➤ Are these roles essential to address the risks?
➤ Are existing laws sufficient?

Bob Harness
Monsanto Company

Nina Fedoroff
Willaman Professor of Life Sciences,
Director Life Sciences Consortium, and
Director Biotechnology Institute
Pennsylvania State University

Marc Lappe
Director
Center for Ethics and Toxics

Margaret Mellon
Agriculture and Biotechnology Program Director
Union of Concerned Scientists

Dennis Gonsalves
Liberty Hyde Bailey Professor
Department of Plant Pathology
Cornell University

Resource Panel – To supplement the panel and answer questions of fact about the coordinated regulatory framework

Janet Anderson
Director
Biopesticides and Pollution Prevention Division
Environmental Protection Agency

James Maryanski
Strategic Manager for Biotechnology, Center for Food Safety and Applied Nutrition
U.S. Food and Drug Administration

Sally McCammon
Science Advisor
Animal and Plant Health Inspection Service
U.S. Department of Agriculture

3:45 – 4:00	Break

4:00 – 4:45	**Voluntary comments from audience at the workshop** What comments, suggestions, or information does the audience have pertaining to the committee's review of the scientific risks/benefits of genetically modified pest protected plants and of the scientific basis for the regulatory review of these plants?

4:45 – 5:00	**Summary Remarks**

Perry Adkisson, Committee Chair

Appendix D

Acronyms

APHIS	Animal and Plant Health Inspection Service
BIO	Biotechnology Industry Organization
Bt	*Bacillus thuringiensis*
CBI	Confidential Business Information
DNA	Deoxyribonucleic Acid
DHHS	Department of Health and Human Services
EPA	Environmental Protection Agency
FDA	Food and Drug Administration
FFDCA	Federal Food, Drug and Cosmetic Act
FPPA	Federal Plant Pest Act
FIFRA	Federal Insecticide, Fungicide, and Rodenticide Act
FONSI	Finding of No Significant Impact
FQPA	Food Quality Protection Act
GMO	Genetically Modified Organism
GMPP	Genetically Modified Pest-Protected
GRAS	Generally Recognized As Safe
HHI	Herfindahl-Hirschman Index
IPM	Integrated Pest Management
MOU	Memorandum of Understanding
NEPA	National Environmental Policy Act
NPDC	Natural Plant Defensive Compounds
NPIRS	National Pesticide Information Retrieval System
OPP	Office of Pesticide Programs
OSTP	Office of Science Technology and Policy
PTGS	Post-transcriptional Gene Silencing
QTL	Quantitative Trait Loci
RAC	Recombinant DNA Advisory Committee
R&D	Research and Development
rDNA	Recombinant DNA
SAP	Scientific Advisory Panel
TSCA	Toxic Substances Control Act
USDA	US Department of Agriculture

Appendix E

Scientific and Common Names

TABLE E.1   Index of Plant Names

Common name	Latin name
alfalfa	*Medicago sativa*
arabidopsis	*Arabidopsis thaliana*
barley	*Hordeum vulgare*
brazil nut	*Bertholletia excelsa*
brussels sprout	*Brassica oleracea* ssp. *gemmifera*
corn	*Zea mays*
cotton	*Gossypium hirsutum*
FLCP (Free-living *Cucurbita pepo*) or Texas gourd	*Cucurbita texana*
maize	*Zea mays*
mlkweed	*Asclepias* sp
oats	*Avena sativa*
papaya	*Carica papaya*
pea	*Pisum sativum*
peanut	*Arachis hypogeae*
potato	*Solanum* spp.
rice	*Oryza sativa*
rye	*Secale cereale* L.
soybean	*Glycine max*
squash	*Cucurbita pepo*
tall wheat grass	*Thinopyrum ponticum*
texas gourd or FLCP	*Cucurbita texana*
tobacco	*Nicotiana tabacum*
tomato	*Lycopersicum esculentum*
wheat, common	*Triticum aestivum* L.
wheat, durum	*Triticum durum* Desf.
wheat, wild	*Triticum monococcum*

TABLE E.2   Index of Plant Diseases

Latin name or Abbreviation	Common Name/Description
Agrobacterium tumefaciens	crown gall bacterium
Cladosporium fulvum	leaf mold fungus
CMV	cucumber mosaic virus
Cochliobolus carbonum	leaf spot fungus
Cochliobolus victoriae	victoria blight (oats)
Erisyphe graminis f. sp. *hordei*	powdery mildew fungus
Erwinia herbicola	bacterial soft rot
Fusarium	ear and stalk rot fungus
Gaeumannomyces graminis	sheath rot fungus
Helminthosporium maydis	southern corn leaf blight and stalk rot fungus
Meloidogyne incognita	root-knot nematode
PRSV	papaya ringspot virus
Pseudomonas syringae	bacterial leaf blight
Pseudocercosporella herpotrichoides	eyespot fungus
Puccinia coronata	crown rust fungus
Puccinia graminis Pers. f. sp. *tritici*	stem rust (or black rust) fungus
Puccinia recondita Rob. Ex Desm. f. sp. *Tritici*	leaf rust (or brown rust) fungus
Puccinia striiformis West	stripe rust (or yellow rust) fungus
TMV	tobacco mosaic virus
WMV2	watermelon mosaic virus-2
ZYMV	zucchini yellow mosaic virus

TABLE E.3   Index of Invertebrates, Plant Pests

Latin name or Abbreviation	Common Name
Campoletis sonorensis	parasitoid of corn earworm
Collembola	springtail
Daphnia	water flea
Helicoverpa zea	corn earworm, cotton bollworm
Heliothis virescens F.	tobacco budworm
Macrosiphum euphorbiae	potato aphid
Pectinophora gossypiella	pink bollworm

Appendix F

Committee Biographical Information

PERRY ADKISSON (CHAIR), CHANCELLOR EMERITUS AND DISTINGUISHED PROFESSOR EMERITUS, TEXAS A&M UNIVERSITY

Dr. Adkisson is known as the father of integrated pest management and was awarded the World Food Prize in 1997 for his significant contributions in plant protection. As Chancellor of the Texas A&M University System, Dr. Adkisson provided extraordinary leadership in a major agricultural state in meeting the needs of state clientele and at the same time developing the university into a world class research institution. He has had a distinguished career as a researcher, educator, and administrator. Dr. Adkisson was elected to the National Academy of Sciences (NAS) in 1979 for his cutting-edge research in entomology and integrated pest management programs. He has served on a number of National Research Council (NRC) boards and commissions including Board on Agriculture and Natural Resources, Food and Nutrition Board, Commission on Life Sciences, and several NAS/NRC ad hoc committees. Dr. Adkisson received his Ph.D. in entomology from Kansas State University.

STANLEY ABRAMSON, MEMBER, ARENT FOX KINTNER PLOTKIN & KAHN, WASHINGTON, DC

Mr. Abramson's law practice is focused on biotechnology, food safety, and environmental law. He works extensively with risk assessment and risk management issues and assists clients with obtaining regulatory approvals for agricultural, industrial and consumer products, including products of genetic engineering. Mr. Abramson represents clients in federal and state enforcement proceedings and defends products before regulatory agencies and the courts. He is the past Associate General Counsel for Pesticides and Toxic Substances at EPA and a principal author of the federal government's Coordinated Framework for Regulation of Biotechnology. Mr. Abramson received his J.D. from Rutgers University.

STEPHEN BAENZIGER, EUGENE W. PRICE PROFESSOR, DEPARTMENT OF AGRONOMY, UNIVERSITY OF NEBRASKA

Dr. Baenziger's research interests focus on cultivar and germplasm development for small grains and improved breeding methodology. New breeding methods include use of tissue culture, genetic engineering, cytogenetic stocks, molecular markers, and statistical designs. Dr. Baenziger has had experience as a plant breeder in industry and with the US Department of Agriculture. He obtained his B.A. in biochemical sciences at Harvard University and his M.S. and Ph.D. in plant breeding and genetics from Purdue University.

FRED BETZ, SENIOR SCIENTIST, JELLINEK, SCHWARTZ & CONNOLLY, ARLINGTON ,VA

Mr. Betz directs the biotechnology and biopesticides practice for this environmental consulting firm. He advises companies in regulatory strategy, provides technical advice on biopesticide matters, and assists with product registrations. These registrations include several of the first genetically engineered plant-pesticides and the first genetically engineered microbial pesticides. Mr. Betz was the principal scientist and regulatory specialist at EPA, where he was responsible for biopesticide risk assessment and biotechnology policy. He received his M.S. in environmental engineering from the University of Florida.

JAMES C. CARRINGTON, PROFESSOR, INSTITUTE OF BIOLOGICAL CHEMISTRY, WASHINGTON STATE UNIVERSITY.

Dr. Carrington is a past NSF Presidential Young Investigator and recipient of the Individual National Research Service Award from NIH. Dr. Carrington's research focuses on viral infection in *Arabidopsis*, host responses to viruses, genetic analysis of RNA virus-host interaction, and activity and transport of viral proteins. He received his Ph.D. in plant pathology from the University of California, Berkeley.

Rebecca Goldburg , Senior Scientist, Environmental Defense, New York, NY

Dr. Goldburg has been responsible for Environmental Defense policy on a number of issues concerning agricultural biotechnology. Dr. Goldburg has authored a number of publications and lectured extensively on food safety, including allergenicity, and on environmental concerns associated with transgenic crops. She received her Ph.D. in ecology from the University of Minnesota.

Fred Gould, William Neal Reynolds Professor, Department of Entomology, North Carolina State University

Dr. Gould's expertise is in the ecological, genetic, and chemical aspects of plant/herbivore interactions. In particular, he is conducting seminal work on resistance management strategies for transgenic crops containing *Bacillus thuringiensis* (Bt) insecticidal genes. Dr. Gould has served on several NRC committees including the Committee on the Future Role of Pesticides in Agriculture and on the Committee on Biological Pest and Pathogen Control. He was involved in the NRC report titled *Ecologically Based Pest Management: New Solutions for a New Century* (1995). Dr. Gould received his Ph.D. from SUNY at Stony Brook.

Ernest Hodgson, William Neal Reynolds Professor of Toxicology, North Carolina State University

Dr. Hodgson's areas of expertise are pesticide metabolism and toxicology. In particular, his lab studies the induction and mechanism of human and rodent cytochrome P450 isoforms in oxidizing agricultural chemicals. Dr. Hodgson has authored well-known basic textbooks of toxicology, *A Textbook of Modern Toxicology* and *An Introduction to Biochemical Toxicology*. He served on the NRC committee which authored the report *Carcinogens and Anticarcinogens in the Human Diet: A Comparison of Naturally Occurring and Synthetic Substances* (1996) and is the editor of the *Journal of Biochemical and Molecular Toxicology*. Dr. Hodgson received his Ph.D. in Biochemistry from Oregon State University.

Tobi Jones, Special Assistant for Special Projects and Public Outreach, California Department of Pesticide Regulation, California Environmental Protection Agency

Dr. Jones is recognized as a national expert in state and federal pesticide regulatory programs, and emerging regulatory issues affecting genetically engineered organisms. She has a strong working knowledge of evolving biotechnology regulation and genetic engineering issues. Dr. Jones is an experienced manager of science-based regulatory programs, serves as the state agency's representative at the national level, and chairs the national committee of state pesticide officials that advises US EPA on federal-state partnership issues. Dr. Jones received her Ph.D. in microbiology from Northwestern University.

MORRIS LEVIN, PROFESSOR, UNIVERSITY OF MARYLAND AND BIOTECHNOLOGY INSTITUTE'S CENTER FOR PUBLIC ISSUES IN BIOTECHNOLOGY

Dr. Levin has extensive experience with risk assessment for genetically engineered organisms both in the federal government and in academia. He has been employed by the Department of Defense and the Department of Interior as an environmental microbiologist. Dr. Levin was in Biotechnology Risk Assessment at the EPA and continues to do risk assessment research at the University of Maryland. He has edited several books including the recent publication *Engineered Organisms in Environmental Settings: Biotechnological and Agricultural Applications* (1996). Dr. Levin received his Ph.D. in Bacteriology and Biophysics from the University of Rhode Island.

ERIK LICHTENBERG, PROFESSOR, DEPARTMENT OF AGRICULTURAL AND RESOURCE ECONOMICS, UNIVERSITY OF MARYLAND

Dr. Lichtenberg has conducted extensive research on productivity and environmental tradeoffs of pesticide regulation, economics of pesticide regulation, and risk-reducing and risk-increasing effects of pesticide regulation. He has served as the Senior Agricultural and Natural Resources Economist on the President's Council of Economic Advisers, as a consultant to the pesticide industry and to EPA, and as an advisor to USDA and EPA. Dr. Lichtenberg received his Ph.D. in agricultural and resource economics from the University of California at Berkeley.

ALLISON SNOW, ASSOCIATE PROFESSOR, DEPARTMENT OF EVOLUTION, ECOLOGY, AND ORGANISMAL BIOLOGY, OHIO STATE UNIVERSITY

Dr. Snow has conducted important research on gene flow and hybridization in several crop-weed systems. Her lab uses molecular techniques to investigate transgene escape to weedy relatives of crops. Dr. Snow's current research focuses on the effects of transgenic insect resistance on herbivory and fitness in wild sunflowers. Dr. Snow has published extensively on the ecological implications of genetically modified crops and has been an associate editor for the journals *Ecology* and *Evolution*. She served on the steering committee for the recent USDA-funded Workshop on the Ecological Effects of Pest Resistance Genes in Managed Ecosystems. Dr. Snow received her Ph.D. in Botany from the University of Massachusetts.

STAFF BIOGRAPHICAL INFORMATION

Jennifer Kuzma, Study Director, serves as Study Director for the NRC Committee on Genetically Modified Pest-Protected Plants and the Standing Committee on Biotechnology, Food and Fiber Production, and the Environment. She has conducted research in the areas of plant biochemistry, molecular biology, and microbiology. Dr. Kuzma was part of a team which identified a novel signal transduction intermediate as a trigger for plant responses to cold, drought, and salinity, and she holds a US patent for the discovery of bacterial isoprene emission. Prior to joining the NRC, she was a AAAS Science Policy Fellow at the USDA where she worked on several microbial food-safety risk assessment projects. She obtained her Ph.D. in Biochemistry from the University of Colorado at Boulder.

Jamie Young, Research Associate is with the National Research Council's Board on Environmental Studies and Toxicology. Her interests include botany, ecology, and natural resources conservation and she is currently nearing completion of her ALB (natural sciences) degree from Harvard University Extension School.

Karen L. Imhof, Project Assistant, joined the Board on Agriculture and Natural Resources in the summer of 1998. Before coming to NAS, Karen worked as a staff assistant in a variety of DC offices. Previous work experience includes the National Wildlife Federation, the Lawyers' Committee for Civil Rights Under Law and the Three Mile Island nuclear facility. Her personal interests include earth and social studies, reading, laughing, and beading.

Derek Sweatt, Project Assistant, joined the Board on Biology in October of 1999. He earned a B.A. in Biology from Colgate University and is currently working towards his M.S. in Environmental Science at Johns Hopkins University.

Index